I0390876

\<Python & Math Series\>

Calculus Story I

with Python

Author

The author studied at Seoul National University (Ph.D.) and currently serves as the director of Nnode LTD. He is interested in analyzing data using R and Pyhton.

He published "Create forecast model for stock with regression analysis and R", "python manual" and "Trigonometry and Limit Story".

sonhs67@gmail.com

datastory1.blogspot.com

Preface

Python is one of the most popular programming languages and is used in many different areas. Unlike other languages, it has a grammar familiar to people's language, so it is easy to learn and has low barriers to application. In particular, sympy, a python module introduced in this book, can represent most theories and expressions of mathematics, thus facilitating the acquisition of concepts as well as complex calculations.

This book mainly uses the sympy module of python to understand the concepts of differential and integral, and introduces various calculations of differential and integral. Derivatives and integrals are used to implicitly denote the meaning of an expression. In order to understand the implications, it is necessary to understand the calculation process of expressions. In order to understand such a meaning, various methods are used in calculus. This book introduces various techniques of calculus and the various mathematical knowledge used in its calculations using python. This course will help you understand mathematical concepts in this area as well as understand and use the python language.

I hope that readers will become more familiar with the math and python language through this book.

Author   Hyun Seok Son

Category

1 Derivative

1.1 Differential concept

Using the limit concept, the slope of the tangent at x=a above the function f(x) is calculated as Eq(1). This slope means the instantaneous rate of change at that point.

$$\lim_{x \to a} \frac{f(x)-f(a)}{x-a} \qquad \text{Eq(1)}$$

From Eq(1), replacing x-a of the denominator with h and approaching h with 0, the expression can be expressed as Eq(2).

$$x-a=h \to \lim_{h \to 0} \frac{f(a+h)-f(a)}{h} \qquad \text{Eq(2)}$$

Eq(2) is called derivative of the function f(x) with respect to x and expressed as f'(x).

Ex) Calculate the derivative of f(x)=2x²-16x+35?

$$f'(x)=\lim_{h \to 0} \frac{2(x+h)^2-16(x+h)+35-(2x^2-16x+35)}{h}$$

$$=\lim_{h \to 0} \frac{2h^2+4xh-16h}{h}$$

$$=\lim_{h \to 0} 2h+4x-16$$

$$=4x-16$$

```
>>> x, h=symbols('x, h')
>>> f=2*x**2-16*x+35;f
2*x**2 - 16*x + 35
>>> f1=f.subs(x, x+h);f1
-16*h - 16*x + 2*(h + x)**2 + 35
>>> f2=f1-f;f2
-16*h - 2*x**2 + 2*(h + x)**2
>>> simplify(f2)
2*h*(h + 2*x - 8)
>>> limit(f2/h, h, 0)
4*x - 16
```

The above calculations can be directly differentiated by applying the diff() function.

```
>>> diff(f, x)
4*x - 16
```

The above procedure can be calculated using functions in the sympy module of python.

1) Converts the character to be used in an expression or function to a symbol object. This conversion uses the symbols() function.

2) Create function expression to use.

3) Create a new expression that adds 'h' to the function.

4) Calculate as Eq(1) above.

These calculations use the sympy module's limit() function.

5) The <u>diff()</u> function of the sympy module can replace steps 3 and 4 above.

Ex) Calculate the derivative of following functions?

1) $g(t) = \dfrac{t}{t+1}$

$$\lim_{h \to 0} \dfrac{\dfrac{t+h}{t+h+1} - \dfrac{t}{t+1}}{h} = \lim_{h \to 0} \dfrac{1}{(t+1)(h+t+1)} = \dfrac{1}{(t+1)^2}$$

```
>>> x, t, h=symbols("x, t, h")
>>> g=t/(t+1);g
t/(t + 1)
>>> g1=g.subs(t, t+h);g1
(h + t)/(h + t + 1)
>>> g2=(g1-g)/h;g2
(-t/(t + 1) + (h + t)/(h + t + 1))/h
>>> g3=together(g2);g3
(-t*(h + t + 1) + (h + t)*(t + 1))/(h*(t + 1)*(h + t + 1))
>>> factor(g3)
1/((t + 1)*(h + t + 1))
>>> dg=limit(g3, h, 0);dg
1/(t**2 + 2*t + 1)
>>> factor(dg)
(t + 1)**(-2)
```

It can be computed by the diff() function.

```
>>> factor(diff(g, t))
(t + 1)**(-2)
```

<Code>

<u>symbols(character)</u>: Specify a character as a symbol.

Expression.<u>subs(a, b)</u>: substitute symbol a for b in expression

<u>together(fraction)</u>: Combine multiple fractions into a single fraction

<u>factor(expression)</u>: factorizing an expression

<u>limit(expression, symbol, value)</u>: calculation of limit of expression

<u>diff(expression)</u>: Differentiate the expression

2) $f(z)=\sqrt{5z-8}$

$$\lim_{h\to 0}\frac{\sqrt{5(z+h)-8}-\sqrt{5z-8}}{h}=\lim_{h\to 0}\frac{(5(z+h)-8)^2-(5z-8)^2}{h(\sqrt{5(z+h)-8}+\sqrt{5z-8})}=\frac{5}{2\sqrt{5z-8}}$$

```
>>> z, h=symbols('z, h')
>>> f=sqrt(5*z-8);f
sqrt(5*z - 8)
```

using limit();

```
>>> f1=(f.subs(z, z+h)-f)/h;f1
(-sqrt(5*z - 8) + sqrt(5*h + 5*z - 8))/h
>>> limit(f1, h, 0)
5/(2*sqrt(5*z - 8))
```

using diff();

```
>>> diff(f, z)
2.5*(5*z - 8)**(-0.5)
```

Ex) f'(0) from f(x)=|x|?

$$\lim_{h \to 0} \frac{|x+h|-|x|}{h} = \lim_{h \to 0} \frac{(x+h)^2 - x^2}{h(|x+h|+|x|)}$$

$$= \lim_{h \to 0} \frac{2xh + h^2}{h(|x+h|+|x|)}$$

$$= \lim_{h \to 0} \frac{2x+h}{|x+h|+|x|}$$

$$= \frac{x}{|x|}$$

$$= \begin{cases} x>0 & 1 \\ x=0 & \text{not exist} \\ x<0 & -1 \end{cases}$$

```
>>> x=symbols('x', real=True)
>>> f=abs(x);f
Abs(x)
>>> diff(f, x).subs(x, 0)
0
>>> dx=diff(f, x);dx
sign(x)
>>> df.subs(x, -3)
-1
>>> df.subs(x,0)
0
>>> df.subs(x, 3)
1
```

If the above function is directly differentiated by the differential formula, it is a constant. Since f'(x) = 1 or -1, there is no change according to x value.

The result by a limit for the derivative of the above function is x/| x|. If x is 0, the denominator is zero and is undefined. However, in the range of the other x, both methods have the same result.

To apply the limit() for the derivative of a fraction, the following limt property is used.

$$\lim_{h \to 0} \frac{g(x)}{f(x)} = \frac{\lim_{h \to 0} g(x)}{\lim_{h \to 0} f(x)}, \qquad f(x) \neq 0$$

The condition of the above properties is that the denominator is non-zero. However, h of denominator can be removed by the factorization of the fraction. (A fraction must be calculated after it has been simplified by factorization or decomposition into partial fractions.) The denominator of the result is |x|, so the limit is not exist at x=0.

```
>>> nom=simplify(f1[0]*(abs(x+h)+abs(x)));nom
h*(h + 2*x)
>>> denom=f1[1]*(abs(x+h)+abs(x));denom
h*(Abs(x) + Abs(h + x))
>>> f2=fraction(nom/denom);f2
(h + 2*x, Abs(x) + Abs(h + x))
```

14

```
>>> limit(f2[0], h,0)
2*x
>>> limit(f2[1],h,0)
x*sign(x) + Abs(x)
>>> f3=limit(f2[0], h,0)/limit(f2[1],h,0);f3
2*x/(x*sign(x) + Abs(x))
>>> f3.subs(x, -3)
-1
>>> f3.subs(x, 3)
1
>>> f3.subs(x, 0)
nan
```

<code>

abs(x): Absolute value of x

sign(x): This function means that the sign changes depending on the scopy of x.

<Definition>

If f '(a) exists at the point x=a, the limit of the function exists and is continuous. Therefore, the function f (x) in a differential interval or point is continuous.

At x=a, f'(x) exist ⇔ continuity

Note that the opposite relationship is not established.

The following is an example of differential calculation using the concept of limit.

$f(x)=x^2$

$$f'(x) = \lim_{h \to 0} \frac{(x+h)^2 - x^2}{h}$$

$$= \lim_{h \to 0} \frac{2xh + h^2}{h}$$

$$= \lim_{h \to 0} 2x + h$$

$$= 2x$$

The above procedure can be generalized by the following differential formula.

$$f(x) = x^n \rightarrow f'(x) = nx^{n-1}$$

The above formula means that the function f is differentiated with respect to the variable x. Therefore, it can be expressed as follows.

$$f'(x) = \lim_{x \to a} \frac{f(x) - f(a)}{x - a} = \frac{df}{dx}\bigg|_{x=a} = f'(a)$$

As shown in the above equation, the derivative is equal to the average rate of change between point a and another point approaching that.

The average rate of change can be interpreted as the instantaneous rate of change at one point since the difference between the points becomes very small as x approaches a.

Ex) About $f(x) = 2x^2 - 16x + 35$

1) The rate of change at x=1?

$f'(x) = 4x - 16 \rightarrow f'(1) = -12$

```
>>> x=symbols('x')
>>> f=2*x**2-16*x+35
>>> dx=diff(f); dx
4*x - 16
>>> dx.subs(x,1)
-12
```

2) The rate of change at x=5?

```
>>> dx.subs(x, 5)
4
```

Figure 1.1 shows the tangent to the position of the function f (x) corresponding to x = 1 and 5.

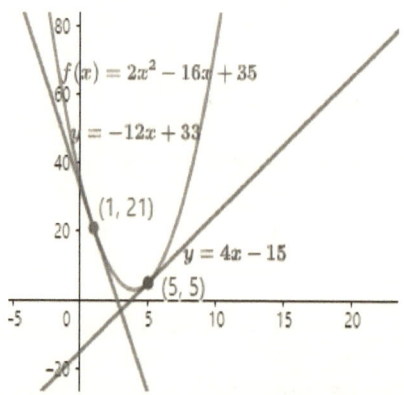

Figure 1.1. Tangent line at x=1, 5 in f(x).

For example, the slope of the tangent line at (1, f (1)) shown in Figure 1.1 is f '(1) which is the derivative value(limit) at that position. This slope can be expressed as the average rate of change between its position and any point approaching it:

$$f'(1)=\frac{f(x)-f(1)}{x-1}$$

From the above equation, the tangent equation is arranged as follows.

y = f'(1)(x-1)+f(1) = -12(x-1)+21 = -12x+33

The result of above procedure is generalized to the formula for tangent equation as like the following equation:

<Tangent Equation>

Tangential equation at x = a

  y=f'(a)(x-a)+f(a)

Since the derivative is the slope of the tangent line, you can decide to increase or decrease the function at that position according to the sign of the derivative value.(See Graph Shape)

In the above case, if x = 1, the function is decremented and at x = 5, the graph is incremented.

Ex) Tangent line of $r(x)=(5x-8)^{1/2}$ at x=3?

```
>>> r=sqrt(5*x-8);r
sqrt(5*x - 8)
```

```
>>> dr=diff(r, x); dr
5/(2*sqrt(5*x - 8))
>>> slope=dr.subs(x, 3); N(slope, 4)
0.9449
```

The function 'tangentLineS' is a user-defined function for calculating the tangent equation.

```
>>> eq=tangentLineS(slope, 3, r.subs(x, 3))
>>> nsimplify(eq, rational=True)
5*sqrt(7)*(x - 3)/14 + sqrt(7)
```

```
<code>
def tangentLineS(slope, a, f_a):
    ... return(slope*(x-a)+f_a)
```

Of the arguments of the above function, slope is the derivative result.

N(x, y): The function indicates any number(x) to the specified number(y) of significant digits.

nsimplify(x): This function makes the coefficient(s) in an expression(x) irreducible fraction(s)

If the position of an object is expressed as a function of f (x), the derivative of the function represents the change in the position, that is, the velocity.

Ex) The position of the object after time t is expressed by the function g(t).

$$g(t)=\frac{t}{t+1}$$

1) Velocity after 10 hours?

```
>>> t=symbols('t')
>>> g=t/(t+1)
>>> dg=diff(g, t); dg
-t/(t + 1)**2 + 1/(t + 1)
>>> dg.subs(t, 10)
1/121
```

2) Could this object stop?

Stopping an object means that the speed is zero. The velocity of this object is dg, which is the derivative of the above function.

```
>>> factor(dg)
(t + 1)**(-2)
```

The derivative result, $1/(t+1)^2$, can not be zero. Therefore, this object can not be stopped.

1.2 Differential properties

1.2.1 Differential formula

<Differential formula>

$$f(x)=x^n \rightarrow \frac{d}{dx} f(x) = f'(x) = nx^{n-1}$$

The above formula is proved using binomial theorem. In this theorem, the condition n> 0, k≥ n is required.

<Binomial theorem>

$$(a+b)^n = \sum_{k=0}^{n} \binom{n}{k} a^{n-k} b^k$$

$$= a^n + \binom{n}{1} a^{n-1}b + \binom{n}{2} a^{n-2}b^2 + \cdots + \binom{n}{n-1} ab^{n-1} + \binom{n}{n} ab^n$$

$$= a^n + na^{n-1}b + \frac{n(n-1)}{2!} a^{n-2}b^2 + \frac{n(n-1)(n-2)}{3!} a^{n-3}b^3 + \cdots + nab^{n-1} + b^n$$

In the above equation, the coefficients of each term are expressed as a combination. The combination is calculated as follows:

$$\binom{n}{k} = \frac{n!}{k!(n-k)!}$$

$$n! = n \cdot (n-1) \cdot \cdots \cdot 2 \cdot 1$$

Applying the limit to the derivative of $f(x)=x^n$ is computed as follows:

$$f'(x)=\lim_{h\to 0}\frac{(x+h)^n-x^n}{h}$$

The right-hand side of the above equation is calculated as follows.

$$\lim_{h\to 0}\frac{(x^n+nx^{n-1}h+\dfrac{n(n-1)}{2!}x^{x-2}h^2+\dfrac{n(n-1)(n-2)}{3!}x^{n-3}h^3+\cdots+nxh^{n-1}+h^n)-x^n}{h}$$

$$=\lim_{h\to 0}\left(nx^{n-1}+\frac{n(n-1)}{2!}x^{x-2}h+\frac{n(n-1)(n-2)}{3!}x^{n-3}h^2+\cdots+nxh^{n-2}+h^{n-1}\right)$$

$$=nx^{n-1}$$

The following characteristics can be applied to the operation of the derivative.

1) Differential of constant

$f(x)=c \rightarrow f'(x)=0$, c: constant

$$f'(x)=\lim_{h\to 0}\frac{f(x+h)-f(x)}{h}=\lim_{h\to 0}\frac{c-c}{h}=0$$

2) $(f(x)\pm g(x))'=f(x)'\pm g(x)'$

3) $(cf(x))'=cf'(x)$ c: any real number

Ex) About f(x) and g(x)

$f(x)=x^2+3x,\ g(x)=x^3+4x+5$

>>> f=x**2+3*x; f

x**2 + 3*x

22

```
>>> g=x**3+4*x+5; g
x**3 + 4*x + 5
```

1) Derivatives for f(x)+g(x)?

```
>>> df=diff(f, x); df
2*x + 3
>>> dg=diff(g, x); dg
3*x**2 + 4
>>> df+dg
3*x**2 + 2*x + 7
>>> diff(f+g, x)
3*x**2 + 2*x + 7
```

2) Derivative of 3f(x)?

```
>>> diff(3*f, x)
6*x + 9
>>> 3*diff(f, x)
6*x + 9
```

Ex) Derivative of various functions?

1) $y = x^{\frac{2}{3}}(2x - x^2)$

$$\rightarrow y' = \frac{d}{dx}\left(2x^{\frac{5}{3}} - x^{\frac{8}{3}}\right) = \frac{10}{3}x^{\frac{2}{3}} - \frac{8}{3}x^{\frac{5}{3}}$$

```
>>> y=x**Rational('2/3')*(2*x-x**2);y
x**(2/3)*(-x**2 + 2*x)
>>> simplify(diff(y, x))
2*x**(2/3)*(-4*x + 5)/3
```

<code>

Rational('2/3'): 2/3 does not calculate and keeps the fraction form.

simplify(equation): Simplify expression

2) $f(x)=15x^{100}-3x^{12}+5x-46$

```
>>> f=15*x**100-3*x**12+5*x-46;f
15*x**100 - 3*x**12 + 5*x - 46
>>> diff(f, x)
1500*x**99 - 36*x**11 + 5
```

3) $t(x)=x^{1/2}+9x^{7/3}-2x^{-2/5}$

```
>>> t=x**0.5+9*x**Rational('7/3')-2*x**Rational('-2/5');t
9*x**(7/3) + x**0.5 - 2/x**(2/5)
>>> diff(t, x)
0.5*x**(-0.5) + 21*x**(4/3) + 4/(5*x**(7/5))
```

Ex) Increasing or decreasing of f(x) at x=-2 ? Tangent line?

$f(x) = 2x^3+300x^{-3}+4$

```
>>> f=2*x**3+300*x**(-3)+4;f
2*x**3 + 4 + 300/x**3
>>> df=diff(f, x); df
6*x**2 - 900/x**4
>>> slope=df.subs(x, -2);slope
-129/4
```

At x=2, the derivative value, the rate of change, is negative, so it decreases. To calculate the tangent expression at this point, the user-defined function tangentLineS is applied as follows:.

```
>>> nsimplify(tangentLineS(slope, 2, f.subs(x, 2)), rational=True)
-129*x/4 + 122
```

Figure 1.2 shows the form of the function f(x) and the tangent at position (2, f (2)).

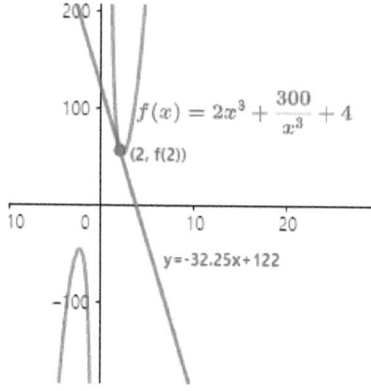

Figure 1.2. f(x)=2x³+300x⁻³+4.

1.2.2 Product, Quotient, and chain Rule

Product Rule

$$(f(x)g(x))' = f'(x)g(x) + f(x)g'(x)$$

The limit expression for the derivative of the product of the two functions is:

$$(f(x)g(x))' = \lim_{h \to 0} \frac{f(x+h)g(x+h) - f(x)g(x)}{h}$$

In above limit, the expressions are rearranged as follows.

$$\frac{f(x+h)g(x+h) - f(x+h)g(x) + f(x+h)g(x) - f(x)g(x)}{h}$$

$$= \frac{f(x+h)(g(x+h) - g(x))}{h} + \frac{g(x)(f(x+h) - f(x))}{h}$$

The above equation is substituted again for the limit case, and it is as follows.

$$(f(x)g(x))' = \lim_{h \to 0} \frac{f(x+h)(g(x+h) - g(x))}{h} + \lim_{h \to 0} \frac{g(x)(f(x+h) - f(x))}{h}$$

$$= \lim_{h \to 0} f(x+h) \lim_{h \to 0} \frac{g(x+h) - g(x)}{h} + \lim_{h \to 0} g(x) \lim_{h \to 0} \frac{f(x+h) - f(x)}{h}$$

$$= f(x)g'(x) + f'(x)g(x)$$

Quotient Rule

<Quotient Rule>

$$\left(\frac{f(x)}{g(x)} \right)' = \frac{f'(x)g(x) - f(x)g'(x)}{g^2(x)}$$

$$\left(\frac{f(x)}{g(x)}\right)' = \lim_{h \to 0} \frac{\dfrac{f(x+h)}{g(x+h)} - \dfrac{f(x)}{g(x)}}{h}$$

$$= \lim_{h \to 0} \frac{1}{h}\left(\frac{f(x+h)}{g(x+h)} - \frac{f(x)}{g(x)}\right)$$

In the above limit expression, rational functions are organized as follows.

$$\frac{g(x)f(x+h) - f(x)g(x+h)}{g(x)g(x+h)}$$

$$= \frac{g(x)f(x+h) - f(x)g(x) + f(x)g(x) - f(x)g(x+h)}{g(x)g(x+h)}$$

$$= \frac{g(x)(f(x+h) - f(x))}{g(x)g(x+h)} - \frac{f(x)(g(x+h) - g(x))}{g(x)g(x+h)}$$

Substituting the above result into the limit expression, it is calculated as follows.

$$\lim_{h \to 0} \frac{1}{h}\left(\frac{g(x)(f(x+h) - f(x))}{g(x)g(x+h)} - \frac{f(x)(g(x+h) - g(x))}{g(x)g(x+h)}\right)$$

$$= \lim_{h \to 0} \frac{1}{g(x)g(x+h)}\left(\frac{g(x)(f(x+h) - f(x))}{h} - \frac{f(x)(g(x+h) - g(x))}{h}\right)$$

$$= \lim_{h \to 0} \frac{1}{g(x)g(x+h)}\left(g(x)\lim_{h \to 0}\frac{f(x+h) - f(x)}{h} - f(x)\lim_{h \to 0}\frac{g(x+h) - g(x)}{h}\right)$$

$$= \frac{1}{g^2(x)}(f'(x)g(x) - f(x)g'(x))$$

Ex) Derivative?

1) The product of f(x) and g(x).

$$f(x)=3x^2,\ g(x)=6x^5\ \rightarrow\ \frac{d}{dx}(f(x)\cdot g(x));$$

$$\frac{d}{dx}(3x^2 6x^5)=\frac{d}{dx}(3x^2)6x^5+3x^2\frac{d}{dx}(6x^5)$$
$$=6x6x^5+3x^2 30x^4$$
$$=36x^6+90x^6=126x^6$$

```
>>> x=symbols('x')
>>> f=3*x**2
>>> g=6*x**5
>>> f*g
 18*x**7
>>> diff((f*g), x)
 126*x**6
>>> diff(f, x)*g+f*diff(g, x)
 126*x**6
```

2) f(x)/g(x)?

$$f(x)=3x^2,\ g(x)=6x^5\ \rightarrow\ \frac{d}{dx}\left(\frac{f(x)}{g(x)}\right);$$

$$\frac{d}{dx}\left(\frac{3x^2}{6x^5}\right)=\frac{d}{dx}\left(\frac{1}{2x^3}\right)=\frac{\frac{d}{dx}(1)\cdot 2x^3-1\cdot\frac{d}{dx}(2x^3)'}{(2x^3)^2}=-\frac{3}{2x^4}$$

```
>>> k=f/g;k
 1/(2*x**3)
>>> diff(k)
 -3/(2*x**4)
>>> (diff(f,x)*g-f*diff(g,x))/g**2
 -3/(2*x**4)
```

3) $y = x^{\frac{2}{3}}(2x-x^2)$?

$$\frac{dy}{dx} = \frac{d}{dx}(x^{\frac{2}{3}}) \cdot (2x-x^2) + x^{\frac{2}{3}} \cdot \frac{d}{dx}(2x-x^2)$$

$$= \frac{2}{3}x^{-\frac{1}{3}}(2x-x^2) + x^{\frac{2}{3}}(2-2x)$$

```
>>> y=x**Rational('2/3')*(2*x-x**2); y
 x**(2/3)*(-x**2 + 2*x)
>>> y1=x**Rational('2/3')
>>> y2=2*x-x**2
>>> [y1, y2]
 [x**(2/3), -x**2 + 2*x]
>>> diff(y1, x)*y2+y1*diff(y2, x)
 x**(2/3)*(-2*x + 2) + 2*(-x**2 + 2*x)/(3*x**(1/3))
>>> diff(y)
 x**(2/3)*(-2*x + 2) + 2*(-x**2 + 2*x)/(3*x**(1/3))
```

User define function for product and quotient rules.

<product rule>

```
def ProductRuleS(f, x, g, y=x):
    return(diff(f, x)*g+f*diff(g, y))
```

<quotient rule>

```
def QuotentRuleS(f, x, g, y= x):
    return((diff(f, x)*g-f*diff(g, y))/g**2)
```

Arguments;

f, g : target functions

x, y : variables in the functions

Ex)   Derivative   of   functions   using   ProductRuleS()   and
QuotentRuleS() ?

1) $f(x)=(6x^3-x)(10-20x)$

>>> f=(6*x**3-x)*(10-20*x)
>>> f1=6*x**3-x
>>> f2=10-20*x
>>> [f, f1, f2]
[(-20*x + 10)*(6*x**3 - x), 6*x**3 - x, -20*x + 10]
>>> ProductRuleS(f1,x, f2)
-120*x**3 + 20*x + (-20*x + 10)*(18*x**2 - 1)
>>> diff(f, x)
-120*x**3 + 20*x + (-20*x + 10)*(18*x**2 - 1)

2) $w(x) = \dfrac{3x+9}{2-x}$?

>>> w=(3*x+9)/(2-x)
>>> w1=3*x+9
>>> w2=2-x
>>> factor(diff(w, x))
15/(x - 2)**2
>>> QuotentRuleS(w1,x,w2,x)
15/(-x + 2)**2

3) $h(x) = \dfrac{4\sqrt{x}}{x^2-2}$?

```
>>> h=(4*sqrt(x))/(x**2-2);h
4*sqrt(x)/(x**2 - 2)
>>> h1=4*sqrt(x); h1
4*sqrt(x)
>>> h2=(x**2-2);h2
x**2 - 2
>>> factor(diff(h, x))
-2*(3*x**2 + 2)/(sqrt(x)*(x**2 - 2)**2)
>>> factor(QuotentRuleS(h1,x,h2))
-2*(3*x**2 + 2)/(sqrt(x)*(x**2 - 2)**2)
>>> factor(diff(h, x))==factor(QuotentRuleS(h1,x,h2, x))
True
```

Ex) The following function is the volume of air injected into the balloon at time t. Air changes in balloon at t = 8?

$$v(t) = \frac{6t^{\frac{1}{3}}}{4t+1}$$

```
>>> t=symbols("t")
>>> v=(6*t**Rational("1/3"))/(4*t+1)
>>> v1=6*t**Rational("1/3")
>>> v2=4*t+1
>>> [v, v1, v2]
 [6*t**(1/3)/(4*t + 1), 6*t**(1/3), 4*t + 1]
>>> dv=diff(v, t);factor(dv)
 -2*(8*t - 1)/(t**(2/3)*(4*t + 1)**2)
>>> dv1=QuotentRuleS(v1, t, v2, t);factor(dv1)#=dv
 -2*(8*t - 1)/(t**(2/3)*(4*t + 1)**2)
```

-7/242

Since the rate of change is negative, the air inside the balloon is decreasing.

For three or more functions, the law of product of derivatives applies as follows.

fgh=(fg)h=f(gh)

⇒ (fgh)'=(fg)'h+(fg)h'=f'gh+fg'h+fgh'

Chain rule

<Chain rule>

The chaining principle applies to the derivative of the resulting function by the combination of two functions f (x) and g (x).

The function R (x) is the result of the synthesis of f (x) and g (x).

R(x)=(f °g)(x)=f(g(x))

The derivation of the composite function is done as follows. This differential process is called a chain rule.

R'(x)=df(x)/dg(x)

$$R'(x)=\frac{df(x)}{dg(x)}\frac{dg(x)}{dx}=f'(g(x))g'(x)$$

Ex) Differentiate R(x)?

$$R(x)=(f)(x)=f(g(x))=\sqrt{5x-8}$$
$$f(g(x))=\sqrt{g(x)}, \quad g(x)=5x-8$$
$$\rightarrow \quad R'(x)=\frac{1}{2}g(x)^{-\frac{1}{2}}g'(x)$$
$$=\frac{1}{2}(5x-8)^{-\frac{1}{2}}5$$
$$=\frac{5}{2\sqrt{5x-8}}$$

```
>>> x, g=symbols("x, g")
>>> f=sqrt(g);f
sqrt(g)
>>> df=diff(f, g);df
1/(2*sqrt(g))
>>> g1=5*x-8;g1
5*x - 8
>>> dg1=diff(g1, x);dg
5
>>> df.subs(g, g1)*dg1
5/(2*sqrt(5*x - 8))
```

Chain rules are built into the algorithm of the diff() function.

```
>>> R=sqrt(5*x-8);R
sqrt(5*x - 8)
>>> dR=diff(R, x);dR
5/(2*sqrt(5*x - 8))
```

Ex) Differentiate the following function?

1) $f(x)=\sin(3x^2+x)$

The above function is a combination of the following two functions.

$f(u)=\sin(u),\ u(x)=3x^2+x$

$\rightarrow f(x)=\sin(3x^2+x)$

$\dfrac{df(x)}{du}=\cos(x),\quad \dfrac{du}{dx}=6x+1$

$\dfrac{df(x)}{dx}=\dfrac{df(x)}{du}\dfrac{du}{dx}$

$\qquad = \cos(u)(6x+1)$

$\qquad = \cos(3x^2+x)(6x+1)$

```
>>> f=sin(3*x**2+x);f
sin(3*x**2 + x)
>>> diff(f, x)
(6*x + 1)*cos(3*x**2 + x)
```

2) $g(x)=\cos^4(x)+\cos(x^4)=(\cos(x))^4+\cos(x^4)=g_1(x)+g_2(x)$

if $u(x)=\cos(x),\ g_1(x)=(u(x))^4$

$\rightarrow g_1'(x)=4(u(x))^3u'(x)=-4(\cos(x))^3\sin(x)$

if $v(x)=x^4,\ g_2(x)=\cos(v(x))$

$\rightarrow g_2'(x)=-\sin(v(x))v'(x)=-\sin(x^4)4x^3$

$\Rightarrow g'(x)=-4(\cos(x))^3\sin(x)-\sin(x^4)4x^3$

In above procedure, (cos(x))'=-sin(x).

```
>>> g=cos(x)**4+cos(x**4);g
cos(x)**4 + cos(x**4)
>>> diff(g, x)
-4*x**3*sin(x**4) - 4*sin(x)*cos(x)**3
```

3) $y=\dfrac{(x^3+4)^5}{(1-2x^2)^3}$

The derivative of the above function applies the quotient rule and the chaining rule.

$$y'=\frac{((x^3+4)^5)'(1-2x^2)^3-(x^3+4)^5((1-2x^2)^3)'}{((1-2x^2)^3)^2}$$

In nominator; $((x^3+4)^5)'=5(x^3+4)^4(3x^2)$

$$((1-2x^2)^3)'=3(1-2x^2)^2(-4x)$$

$$\rightarrow y'=\frac{1}{(1-2x^2)^6}(5(x^3+4)^4(3x^2)(1-2x^2)^3-(x^3+4)^53(1-2x^2)^2(-4x))$$

$$=\frac{1}{(1-2x^2)^4}(3x(x^3+4)^4(5x-6x^3+16))$$

```
>>> f=(x**3+4)**5/(1-2*x**2)**3;f
(x**3 + 4)**5/(-2*x**2 + 1)**3
>>> factor(diff(f, x))
-3*x*(x**3 + 4)**4*(6*x**3 - 5*x - 16)/(2*x**2 - 1)**4
```

1.2.3 Intermediate value theorem

<Intermediate value theorem>

If the function f(x) is continuous in the interval [a, b] and f(a)<M<f(b), there exists a value c in the interval that satisfies the following conditions.

1) a<c<b

2) f(c)=M

3) If conditions 1) and 2) are satisfied, f(x) in the open interval (a, b) can be differentiated and the tangent equation in c can be

expressed as follows.

$$f'(c)=\frac{f(b)-f(a)}{b-a}$$
$$\leftrightarrow f(b)-f(a)=f'(c)(b-a)$$
$$\leftrightarrow f(b)=f'(c)(b-a)+f(a)$$

The median value theory does not determine the c value satisfying the above 3), but just determines whether c exists or not.

As shown in Figure 1.3, there is a point c between the interval a and b where the function can be differentiated. The slope of tangent line at c is equal to that of scant line between a and b.

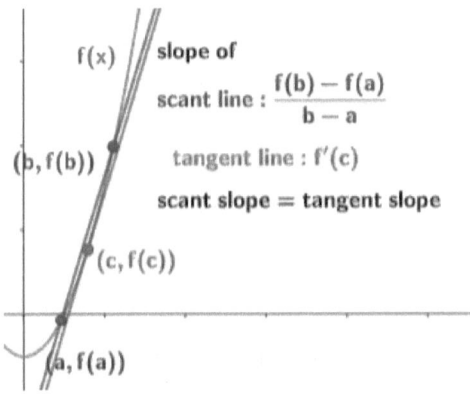

Figure 1.3. Intermediate value theorem.

Ex) In the following function, is the point c which satisfies intermediate value theorem?

36

$f(x)=x^3+2x^2-x,\ [-1, 2]$

```
>>> x=symbols("x")
x**3 + 2*x**2 - x
>>> f=x**3+2*x**2-x; f
x**3 + 2*x**2 - x
>>> df=diff(f, x);df
3*x**2 + 4*x - 1
>>> slope=(f.subs(x, 2)-f.subs(x, -1))/(2--1);slope
4
```

Any point with a derivative value (tangent slope) of 4 is determined between intervals (-1, 2).

$f'(c)=4 \rightarrow 3x^2+4x-1=4$

Use the solve() function to calculate the solution of the above equation.

```
>>> sol=solve(df-slope, x);sol
[-2/3 + sqrt(19)/3, -sqrt(19)/3 - 2/3]
>>> sol1=[i.evalf(4) for i in sol];sol1
[0.7863, -2.120]
```

The solution included in the specified interval is 0.7863.

<code>

<u>solve(equation, symbol)</u>: Calculates the solution of the equation.

1.3 Differentiation of various functions

1.3.1 Trigonometric function

The following are the basic formulas used to derive the various derivative formulas of trigonometric functions.

$$\lim_{\theta \to 0} \frac{\sin(\theta)}{\theta} = 1 \qquad \text{Eq(1)}$$

$$\lim_{\theta \to 0} \frac{\cos(\theta)-1}{\theta} = 0 \qquad \text{Eq(2)}$$

<Proof of Eq(1)>

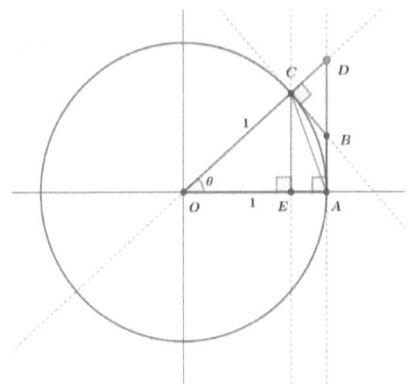

Figure 1.4. Triangle in unit circle.

Figure 1.4 is based on the tangent line between two distal points, C and A. To simplify the proof process, the range and condition of θ are assumed to be $0 \le θ \le π / 2$ and $θ → 0$.

arc AC<|AB|+|BC|　(1)

|BC|<|BD|, |AB|+|BD|=|AD| Therefore, Eq(1) can be modified as follows.

arc AC<|AB|+|BC| <|AB|+|BD|=|AD|　(2)

The tan(θ) in Figure 1.4 is defined as follows.

$$tan(\theta) = \frac{|AD|}{|AO|} \rightarrow |AD| = tan(\theta)|AO| = tan(\theta)$$

∵ |AO|=1

Therefore, (2) is corrected again by the following equation (3).

arc AC < |AD|= tan(θ)|AO|=tan(θ)　(3)

The length of the arc can be expressed in radians by applying the relationship between the circle's angle and its length.

360 : θ = 2πr : arc AC,　r= radius

arc AC=θr=θ|OA|=θ

(4) is the result of substituting the above result into (3).

$$\text{arc } AC = \theta < \tan(\theta) = \frac{\sin(\theta)}{\cos(\theta)}$$

$$\rightarrow \cos(\theta) < \frac{\sin(\theta)}{\theta} \qquad (4)$$

$$|CE| = \sin(\theta) < |AC| < \text{arc } AC = \theta \qquad (5)$$

$$\sin(\theta) < |AC| < \theta \rightarrow \frac{\sin(\theta)}{\theta} < 1 \qquad (6)$$

$$\text{From}(4), \cos(\theta) < \frac{\sin(\theta)}{\theta} < 1 \qquad (7)$$

$$\lim_{\theta \to 0} \cos(\theta) = 1$$

$$\therefore \lim_{\theta \to 0} \frac{\sin(\theta)}{\theta} = 1$$

Eq. 2 is proven as follows.

$$\lim_{h \to 0} \frac{\cos(\theta)-1}{\theta} = \lim_{h \to 0} \frac{(\cos(\theta)-1)(\cos(\theta)+1)}{\theta(\cos(\theta)+1)}$$

$$= \lim_{h \to 0} \frac{\cos^2(\theta)-1}{\theta(\cos(\theta)+1)}$$

$$= \lim_{h \to 0} \frac{-\sin^2(\theta)}{\theta(\cos(\theta)+1)}$$

$$= \lim_{h \to 0} \frac{-\sin(\theta)}{\theta} \lim_{h \to 0} \frac{\sin(\theta)}{\cos(\theta)+1}$$

$$= -1 \cdot \frac{0}{2} = 0$$

The following trigonometric formula is applied in the above procedure.

$$\sin^2(\theta) + \cos^2(\theta) = 1 \rightarrow \cos^2(\theta)-1 = -\sin^2(\theta)$$

Calculating the following expression from the above procedure with sympy's limit() function yields the same result.

$$\lim_{h \to 0} \frac{\sin(\theta)}{\cos(\theta)+1}$$

```
>>> th=symbols("th")
>>> limit(-sin(th)/(cos(th)+1), th, 0)
0
```

<Code>

The sympy module has functions for calculating the value of a trigonometric function. Same as the usual trigonometric form, but the value that should be passed as an argument must be a radian, not an angle.

Ex) Calculate the limit of the following trigonometric functions?

1) $\lim\limits_{\theta \to 0} \dfrac{\sin(\theta)}{6\theta} = \dfrac{1}{6} \lim\limits_{\theta \to 0} \dfrac{\sin(\theta)}{\theta} = \dfrac{1}{6}$

```
>>> th=symbols("th")
>>> limit(sin(th)/(6*th), t, 0)
1/6
```

2) $\lim\limits_{\theta \to 0} \dfrac{\sin(6\theta)}{6\theta}$

$\to x=6\theta,$

$6 \lim\limits_{\theta \to 0} \dfrac{\sin(6\theta)}{6\theta} = \lim\limits_{x \to 0} \dfrac{\sin(x)}{x} = 6$

```
>>> limit(sin(6*t)/(t), t, 0)
6
>>> limit(t/sin(7*t), t, 0)
```

3) $\lim\limits_{x\to 0} \dfrac{\sin(3x)}{\sin(8x)}$

if x=3x, or 8x, x→0 ⟹ 3x, or 8x→0

$\lim\limits_{x\to 0} \dfrac{\sin(3x)}{x} \quad \dfrac{x}{\sin(8x)}$

$= \lim\limits_{3x\to 0} \dfrac{3\sin(3x)}{3x} \quad \lim\limits_{8x\to 0} \dfrac{8x}{8\sin(8x)} = \dfrac{3}{8}$

>>> f=(sin(3*x))/(sin(8*x));f

sin(3*x)/sin(8*x)

>>> limit(f, x, 0)

3/8

4) $\lim\limits_{x\to 4} \dfrac{\sin(x-4)}{x-4}$

Replacing x-4 with t, x→4 ⟹ t→0

$\lim\limits_{t\to 0} \dfrac{\sin(t)}{t} = 1$

>>> x=symbols("x")

>>> f=(sin(x-4))/(x-4);f

sin(x - 4)/(x - 4)

>>> limit(f, x, 4)

1

5) $\lim\limits_{\theta\to 0} \dfrac{\cos(2\theta)-1}{\theta}$

$\lim\limits_{\theta\to 0} \dfrac{2(\cos(2\theta)-1)}{2\theta} = 2 \cdot 0 = 0$

>>> z=symbols("z")

42

```
>>> f=(cos(2*z)-1)/z;f
(cos(2*z) - 1)/z
>>> limit(f, z, 0)
0
```

The derivative formula of the trigonometric function is as follows.

<Trigonometry Differential formula>

$(\sin(x))'=\cos(x)$ $\qquad$ $(\cos(x))'=-\sin(x)$

$(\tan(x))'=\sec^2(x)$

$(\csc(x))' = -\csc(x)\cot(x)$ $\quad$ $(\sec(x))' = \sec(x)\tan(x)$

$(\cot(x))' = -\csc^2(x)$

$$\frac{d}{dx}\sin(x)=\lim_{h\to0}\frac{\sin(x+h)-\sin(x)}{h}$$

<sum formula of trigometry>

$$\sin(x+h)=\sin(x)\cos(h)+\cos(x)\sin(h)$$

$$\to\lim_{h\to0}\frac{\sin(x)\cos(h)+\cos(x)\sin(h)-\sin(x)}{h}$$

$$=\sin(x)\lim_{h\to0}\frac{\cos(h)-1}{h}+\cos(x)\lim_{h\to0}\frac{\sin(h)}{h}$$

$$=\cos(x)$$

$$\frac{d}{dx}\cos(x)=\lim_{h\to0}\frac{\cos(x+h)-\cos(x)}{h}$$

<sum formula of trigometry>

$$\cos(x+h)=\cos(x)\cos(h)-\sin(x)\sin(h)$$

$$\to\lim_{h\to0}\frac{\cos(x)\cos(h)-\sin(x)\sin(h)-\cos(x)}{h}$$

$$= -\sin(x)\lim_{h\to0}\frac{\sin(h)}{h}+\cos(x)\lim_{h\to0}\frac{\cos(h)-1}{h}$$

$$= -\sin(x)$$

$$\frac{d}{dx}\tan(x) = \frac{d}{dx}\left(\frac{\sin(x)}{\cos(x)}\right)$$

$$= \frac{\cos(x)\cos(x) - \sin(x)(-\sin(x))}{\cos^2(x)}$$

$$= \frac{1}{\cos^2(x)}$$

$$= \sec^2(x)$$

Ex) Differentiate the following?

1) $y = 5\sin(x)\cos(x) + 4\csc(x)$

$y' = 5(\sin'(x)\cos(x) + \sin(x)\cos'(x)) + 4\csc'(x) = 5(\cos^2(x) - \sin^2(x)) - 4\csc(x)\cot(x)$

```
>>> x=symbols("x")
>>> f=5*sin(x)*cos(x)+4*csc(x);f
5*sin(x)*cos(x) + 4*csc(x)
>>> diff(f, x)
-5*sin(x)**2 + 5*cos(x)**2 - 4*cot(x)*csc(x)
```

2) $f(x)$

$$f(x) = \frac{\sin(x)}{3 - 2\cos(x)}$$

$$f'(x) = \frac{\cos(x)(3 - 2\cos(x)) - \sin(x)(2\sin(x))}{(3 - 2\cos(x))^2}$$

$$= \frac{3\cos(x) - 2}{(3 - 2\cos(x))^2}$$

```
>>> x=symbols('x')
>>> f=sin(x)/(3-2*cos(x))
>>> diff(f, x)
cos(x)/(-2*cos(x) + 3) - 2*sin(x)**2/(-2*cos(x) + 3)**2
```

44

```
>>> factor(diff(f, x))
-(2*sin(x)**2 + 2*cos(x)**2 - 3*cos(x))/(2*cos(x) - 3)**2
>>> simplify(factor(diff(f, x)))
(3*cos(x) - 2)/(2*cos(x) - 3)**2
```

Ex) The following function is the balance of the bank account per year(t). When does the balance of the account increase for 10 years after opening the account?

$P(t) = 500+100\cos(t)-150\sin(t)$

The rate of change of the balance can be expressed as a derivative of the balance function.

```
>>> t=symbols('t')
>>> p=500+100*cos(t)-150*sin(t);p
-150*sin(t) + 100*cos(t) + 500
>>> dp=diff(p, t); dp
-100*sin(t) - 150*cos(t)
```

The increment or decrement of a function can be estimated by the sign of the differentiated function. Therefore, the sign of the derivative function 'dp' should be investigated.

For this investigation, we will create a program that computes the derivative function values at various points in time from the opening of the initial account to 11 years, and returns the derivative values at the point when the sign of the value is switched.

```
>>> val=[dp.subs(t, 0).evalf(4)];val  #P'(0)
[-150.0]
>>> re={}
```

The period from the beginning to the 11th is divided into 100 sections.

```
>>> rng=np.linspace(0, 11, 100)
>>> n=1
>>> for i in rng:
...     val.append(dp.subs(t, i).evalf(4))
...         if val[n-1]*val[n]<0:
...             re[round(i, 1)]=val[n]
...         n=n+1

>>> re
{2.2: 7.426, 5.4: -17.93, 8.5: 28.37}
>>> re1=pd.DataFrame([re.keys(), re.values()]).T
>>> re1
    0    1
0  2.2  7.426
1  5.4  -17.93
2  8.6  28.37
```

According to the above results, the increase / decrease of differential value is as follows.

0 to 2.2 year: Decrease

2.2~5.4 year: increase

5.4~8.6 year: Decrease

8.6 ~ 11 year: increase

Therefore, for the first time in 2.2 years, the balance starts to increase.

<code>

np.linspace(begin, end, size): Creates a sequence of the specified size at regular intervals from start to end.

object.subs(x, y): substituting y for x in the object

object.append(x): adds a new element(x) to the end of the list object.

The for loop and if conditional statements are used only to indicate results that meet certain conditions.

pd.DataFrame(): Returns the result in the form of a table.

1.3.2 Inverse trigonometric function

A function whose domain and range are exchanged is called an inverse function. That is, if f(x)=y → g(y)=x, these two functions have an inverse function relationship.

Differential formula of inverse function

<Formula of Inverse function>

$$g'(x) = \frac{1}{f'(g(x))}$$

If the inverse function of y=f(x) is g(x), f(g(x))=x. Therefore, the derivative of the function is:

$$f'(g(x))g'(x)=1 \rightarrow g'(x)=\frac{1}{f'(g(x))}$$

Ex) Differentiate the inverse function of the following function?

1) $f(x)$: $y=(x+1)^2 \rightarrow f^{-1}(x)=g(x)$: $x=\sqrt{y}-1$

$$1dx=\frac{1}{2\sqrt{y}}dy=\frac{1}{2(x+1)}dy$$

$$\therefore \frac{dx}{dy}=\frac{1}{2(x+1)}=\frac{1}{f'(x)}=\frac{1}{f'(g(x))}$$

$$\because f(g(x))=(\sqrt{y}-1+1)^2=y=f(x)$$

2) $f(x)=y=x^3 \rightarrow g(x)=x=\sqrt[3]{y}$

$$dx=\frac{1}{3\sqrt[3]{y^2}}dy \rightarrow \frac{dx}{dy}=\frac{1}{3\sqrt[3]{y^2}}=\frac{1}{3x^2}=g'(x)$$

$$g'(x)=\frac{1}{f'(g(x))}=\frac{1}{f'(x)}=\frac{1}{3x^2}$$

The inverse of the trigonometric function is defined as:

<Definition>

$y = \sin^{-1}(x) \Leftrightarrow x = \sin(y)$: $\sin^{-1}(x) = \arcsin(x)$

$y = \cos^{-1}(x) \Leftrightarrow x = \cos(y)$: $\cos^{-1}(x) = \arccos(x)$

$y = \tan^{-1}(x) \Leftrightarrow x = \tan(y)$: $\tan^{-1}(x) = \arctan(x)$

The following relationship is established from the definition of the inverse function.

$\cos(\cos^{-1}(x)) = x$

$\sin(\sin^{-1}(x)) = x$

$\tan(\tan^{-1}(x)) = x$

Therefore, when applied to the derivative formula of the inverse function, the derivative formula of each inverse trigonometry is as follows.

$$\frac{d}{dx}\sin^{-1}(x) = \frac{1}{\sqrt{1-x^2}} \qquad\qquad \frac{d}{dx}\cos^{-1}(x) = -\frac{1}{\sqrt{1-x^2}}$$

$$\frac{d}{dx}\tan^{-1}(x) = \frac{1}{x^2+1}$$

$$\frac{d}{dx}\csc^{-1}(x) = -\frac{1}{|x|\sqrt{1-x^2}} \qquad\qquad \frac{d}{dx}\sec^{-1}(x) = \frac{1}{|x|\sqrt{1-x^2}}$$

$$\frac{d}{dx}\cot^{-1}(x) = -\frac{1}{x^2+1}$$

The calculation of the inverse trigonometric function from the sympy module is performed as a function that connects the trigonometric function with the prefix "a".

$$\sin^{-1}(x) = \mathrm{asin}(x) \qquad \cos^{-1}(x) = \mathrm{acos}(x)$$
$$\tan^{-1}(x) = \mathrm{atan}(x) \qquad \csc^{-1}(x) = \mathrm{acsc}(x)$$
$$\sec^{-1}(x) = \mathrm{asec}(x) \qquad \cot^{-1}(x) = \mathrm{acot}(x)$$

Differentiate of $y = \sin^{-1}(x)$.

$y=\sin^{-1}(x) \rightarrow x=\sin(y)$

From $\sin^2(y)+\cos^2(y)=1$

$\rightarrow \cos^2(y)=\cos^2(\sin^{-1}(x))=1-\sin^2(y)=1-x^2$

$\rightarrow \cos(\sin^{-1}(x))=\sqrt{1-x^2}$

$\dfrac{d}{dx}\cos(\sin^{-1}(x))=-\sin(\sin^{-1}(x))\dfrac{d}{dx}\sin^{-1}(x)=\dfrac{-2x}{2\sqrt{1-x^2}}$

$\rightarrow -x\dfrac{d}{dx}\sin^{-1}(x)=\dfrac{-2x}{2\sqrt{1-x^2}}$

$\therefore \dfrac{d}{dx}\sin^{-1}(x)=\dfrac{1}{\sqrt{1-x^2}}$

```
>>> x=symbols("x")
>>> g=asin(x);g
asin(x)
>>> dg=diff(g, x);dg #dg=g'(x)
1/sqrt(-x**2 + 1)
>>> f=sin(x);f
sin(x)
>>> df=diff(f, x);df
cos(x)
>>> 1/(df.subs(x, g))#1/[f'(g(x))]
1/sqrt(-x**2 + 1)
```

Differentiate of $y = \cos^{-1}(x)$.

$y = \cos^{-1}(x) \rightarrow x = \cos(y)$

From $\sin^2(y) + \cos^2(y) = 1$

$\rightarrow \sin^2(y) = \sin^2(\cos^{-1}(x)) = 1 - \cos^2(y) = 1 - x^2$

$\rightarrow \sin(\cos^{-1}(x)) = \sqrt{1-x^2}$

$\dfrac{d}{dx}\sin(\cos^{-1}(x)) = \cos(\sin^{-1}(x))\dfrac{d}{dx}\cos^{-1}(x) = \dfrac{-2x}{2\sqrt{1-x^2}}$

$\rightarrow x\dfrac{d}{dx}\sin^{-1}(x) = \dfrac{-2x}{2\sqrt{1-x^2}}$

$\therefore \dfrac{d}{dx}\sin^{-1}(x) = \dfrac{-1}{\sqrt{1-x^2}}$

```
>>> g=acos(x);g
acos(x)
>>> dg=diff(g, x);dg #dg=g'(x)
-1/sqrt(-x**2 + 1)
>>> f=cos(x);f
cos(x)
>>> df=diff(f, x);df
-sin(x)
>>> 1/(df.subs(x, g))#1/[f'(g(x))]
-1/sqrt(-x**2 + 1)
```

Differentiate of $y=\tan^{-1}(x)$.

$y=\tan^{-1}(x) \rightarrow x=\tan(y)$

From $\sin^2(y)+\cos^2(y)=1$

$$\rightarrow \frac{\sin^2(y)}{\cos^2(y)}+1=\frac{1}{\cos^2(y)}$$

$\rightarrow \tan^2(y)+1=\sec^2(y)$

$\rightarrow \tan^2(\tan^{-1}(x))+1=\sec^2(\tan^{-1}(x))$

$\rightarrow \sec^2(\tan^{-1}(x))=x^2+1$

$\rightarrow \sec(\tan^{-1}(x))=\sqrt{x^2+1}$

$\because \tan^2(x)=(\tan(x))^2$

Using $\sec'(x)=\sec(x)\tan(x)$,

$$\frac{d}{dx}\sec(\tan^{-1}(x))=\sec(\tan^{-1}(x))\tan(\tan^{-1}(x))\frac{d}{dx}\tan^{-1}(x)=\frac{2x}{2\sqrt{x^2+1}}$$

$$\rightarrow \sqrt{x^2+1} \; x \frac{d}{dx}\tan^{-1}(x)=\frac{2x}{2\sqrt{x^2+1}}$$

$$\rightarrow \frac{d}{dx}\tan^{-1}(x)=\frac{1}{x^2+1}$$

```
>>> g=atan(x);g
atan(x)
>>> dg=diff(g, x);dg #dg=g'(x)
1/(x**2 + 1)
>>> f=tan(x);f
tan(x)
>>> df=diff(f, x);df
tan(x)**2 + 1
>>> 1/(df.subs(x, g))#1/[f'(g(x))]
1/(x**2 + 1)
```

Ex) Differentiate following functions?

52

1) $f(x) = 4\cos^{-1}(x) - 10\tan^{-1}(x)$

```
>>> x=symbols("x")
>>> f=4*acos(x)-10*atan(x);f
4*acos(x) - 10*atan(x)
>>> diff(f, x)
-10/(x**2 + 1) - 4/sqrt(-x**2 + 1)
```

2) $y = \sqrt{z}\,\sin^{-1}(z)$

```
>>> z=symbols("z")
>>> y=z**0.5*asin(z);y
z**0.5*asin(z)
>>> diff(y, z)
0.5*z**(-0.5)*asin(z) + z**0.5/sqrt(-z**2 + 1)
```

1.3.3 Exponential and Logarithmic functions

Exponential function

The derivative of the exponential function can be expanded based on Euler's number, e. So let's first look at the general definition of this Euler number.

<Definition>

$$1) \ e = \lim_{n \to \infty} \left(1 + \frac{1}{n}\right)^n$$

$$2) \ e > 0, \quad \lim_{h \to 0} \frac{e^h - 1}{h} = 1$$

$$3) \ e = \sum_{n=0}^{\infty} \frac{1}{n!}$$

```
>>> n, h=symbols("n, h")
>>> limit((1+1/n)**n, n, oo)
E
>>> exp(1)#E=Euler's number(e) at sympy module
E
>>> limit((exp(h)-1)/h, h, 0)
1
>>> Sum(1/factorial(x), (x, 0, oo))
Sum(1/factorial(x), (x, 0, oo))
>>> Sum(1/factorial(n), (n, 0, oo)).doit()
E #=e or exp
```

\<code\>

Sum(f, (var, dw, up)): Create the object of series.

object.doit(): Calculate a sympy object(ecpression) such as Sum().

factorial(num): Calculate a factorial.

In general, a constant can be expressed using Euler's number and natural logarithm as follows:

a=exp(ln(a))

```
>>> a=symbols('a')
```

```
>>> exp(ln(a))

a
```

Figure 1.5 shows a typical exponential function.

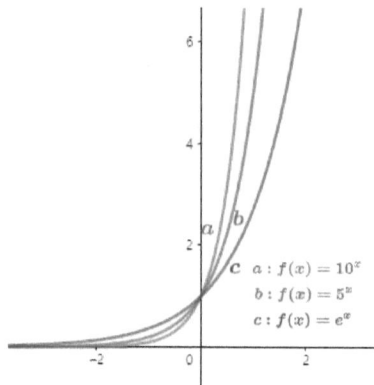

Figure 1.5. Exponential Function.

As shown in Figure 1.5, the value of the exponential function f(x) is positive and approaches f(0) → 1 as it progresses from x → 0.

In general, the exponential function(f(x)) can be expressed in the same format as exp(ln(a)), as follows:

f(x)=a^x=exp(ln(a^x))=exp(x ln(a))

The derivative of the exponential function proceeds as follows.

1) The function is regenerated as a composite function by applying the substitution method.

2) The composite function can be differentiated by the chain law.

$$f(x)=e^{x\ln(a)} \rightarrow f(u)=e^{u},\ u(x)=x\ln(a)$$

$$f'(u)=\lim_{h\to 0}\frac{e^{u+h}-e^{u}}{h}=e^{u}\lim_{h\to 0}\frac{e^{h}-1}{h}=e^{u}$$

$$u'(x)=\ln(a)$$

$$\rightarrow f'(u(x))=\frac{d}{du}f(u)\frac{du}{dx}$$

$$=e^{u}\ln(a)$$
$$=e^{x\ln(a)}\ln(a)$$
$$=a^{x}\ln(a)$$

As a result, the derivative of the natural log is summarized as follows.

$$f(x)=ax \Rightarrow f'(x)=a^{x}\ln(a)$$

```
>>> x, a=symbols('x, a')
>>> f=a**x
>>> diff(f, x)
a**x*log(a)
```

Logarithm function

The log function is the inverse function of the exponential function.

The domain and range $f(x)=e^{x}$ are x and f(x), respectively. The domain and range are exchanged in inverse function.

$$x=b^{y} \rightarrow y=\log_{b}x$$

In the above equation, b is base.

b = 10: Common log, log(a)

b = e (Euler number): natural logarithm, ln(a)

Also, since the range of the exponential function is positive, the domain function of the log function, which is an inverse function of the exponential function, is positive.

In the sympy module, both the commercial and natural logs are represented by the log() function, which is distinguished by the arguments passed to the function.

The derivative of the log function is calculated using the inverse function relationship between the exponential function and the log function.

$y=e^x \rightarrow x=\ln(y)$

$dx=(\ln(y))dy$

$\rightarrow \dfrac{dx}{dy}=\dfrac{1}{\ln(y)}=\dfrac{1}{\ln(e^x)}=\dfrac{1}{x}$

```
>>> x=symbols("x")
>>> f=exp(x);f
exp(x)
>>> g=log(x);g
log(x)
>>> df=diff(f, x); df
exp(x)
>>> 1/(df.subs(x, g))
1/x
>>> diff(g, x)
1/x
```

From above result, The derivative of $\log_a x$ is follows.

$$\ln_a x = \frac{\ln(x)}{\ln(a)}$$

$$\frac{d}{dx}\left(\frac{\ln(x)}{\ln(a)}\right) = \frac{1}{\ln(a)}\frac{d}{dx}(\ln(x)) = \frac{1}{x\ln(a)}$$

The derivative formula of exponent and logarithmic function is as follows.

<Derivative formula of exponent and logarithmic function>

$$\frac{d}{dx}e^x = e^x \qquad \frac{d}{dx}a^x = a^x\ln(a)$$

$$\frac{d}{dx}\ln(x) = \frac{1}{x} \qquad \frac{d}{dx}\ln_a(x) = \frac{1}{x\ln(a)}$$

Ex) Differentiate follows?

1) g(x)=4x-5ln₉x

g'(x)=4xln(4)-5/(x ln(9))

```
>>> x=symbols("x")
>>> g=4**x-5*log(x, 9);g
4**x - 5*log(x)/log(9)
>>> diff(g, x)
4**x*log(4) - 5/(x*log(9))
```

2) $y = \dfrac{5e^x}{3e^x + 1}$

$y' = \dfrac{(5e^x)'(3e^x+1)-(5e^x)(3e^x+1)'}{(3e^x+1)^2}$

$ = \dfrac{5e^x3e^x+5e^x-5e^x3e^x}{(3e^x+1)^2}$

$ = \dfrac{5e^x}{(3e^x+1)^2}$

```
>>> y=(5*exp(x))/(3*exp(x)+1);g
5*exp(x)/(3*exp(x) + 1)
>>> factor(diff(y, x))
5*exp(x)/(3*exp(x) + 1)**2
```

Ex) The movement of an object along time t follows the distance function S(t). Can this object stop??

s(t)=t·exp(t)

The differentiation of the above distance function represents the speed. Therefore, when the differential result becomes 0, the movement of the object becomes zero.

s'(t)=exp(t)+t·exp(t)=(1+t)·exp(t)=0

```
>>> t=symbols("t")
>>> s=t*exp(t);s
t*exp(t)
>>> ds=diff(s, t);ds
t*exp(t) + exp(t)
>>>solve(Eq(ds, 0), t)
[-1]
```

At t=-1, the object stops. However, since the time variable t is not a negative number, this object can not be stopped.

  <code>

  solve(): Calculate the solution(s) of equation(s).

1.3.4 Hyperbolic function

The various hyperbolic functions can be represented by the combination of two exponential functions ex and e-x.

<Hyperbolic functions>

$$\sinh(x) = \frac{e^x - e^{-x}}{2}$$

$$\cosh(x) = \frac{e^x + e^{-x}}{2}$$

$$\tanh(x) = \frac{\sinh(x)}{\cosh(x)} = \frac{e^x - e^{-x}}{e^x + e^{-x}}$$

$$\operatorname{csch}(x) = \frac{1}{\sinh(x)} = \frac{2}{e^x - e^{-x}}$$

$$\operatorname{sech}(x) = \frac{1}{\cosh(x)} = \frac{2}{e^x + e^{-x}}$$

$$\coth(x) = \frac{1}{\tanh(x)} = \frac{e^x + e^{-x}}{e^x - e^{-x}}$$

The hyperbolic function has the following computational properties:

$$\sinh(-x)=\frac{e^{-x}-e^{x}}{2}=-\frac{e^{x}-e^{-x}}{2}=-\sinh(x)$$

$$\cosh(-x)=\frac{e^{-x}+e^{x}}{2}=\frac{e^{x}+e^{-x}}{2}=\cosh(x)$$

$$\cosh^{2}(x)-\sinh^{2}(x)=\left(\frac{e^{x}+e^{-x}}{2}\right)^{2}-\left(\frac{e^{x}-e^{-x}}{2}\right)^{2}=\frac{4e^{x}e^{-x}}{4}=1$$

$$1-\tanh^{2}(x)=1-\left(\frac{e^{x}-e^{-x}}{e^{x}+e^{-x}}\right)^{2}=\frac{(e^{x}+e^{-x})^{2}-(e^{x}-e^{-x})^{2}}{(e^{x}+e^{-x})^{2}}=\left(\frac{2}{e^{x}+e^{-x}}\right)^{2}=\operatorname{sech}^{2}(x)$$

Figure 1.6 is a graph of various hyperbolic functions.

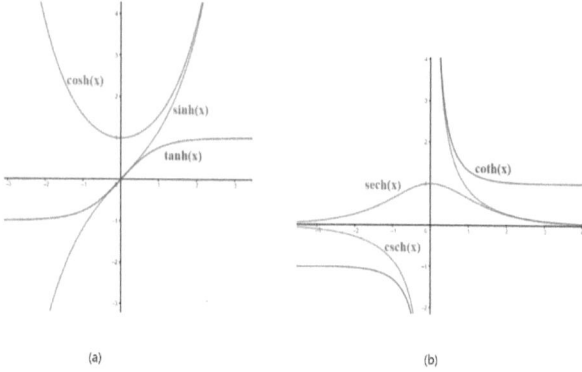

(a)   (b)

Figure 1.6. Various hyperbolic functions.

The derivative of the hyperbolic function generated by the combination of exponential functions is easily derived as a derivative of the exponential function. Typically, the derivative of tanh(x) can be derived as:

61

$$\frac{d}{dx}\tanh(x) = \frac{d}{dx}\frac{\operatorname{sonh}(x)}{\cosh(x)}$$

$$= \frac{d}{dx}\frac{e^x-e^{-x}}{e^x+e^{-x}}$$

$$= \frac{(e^x-e^{-x})'(e^x+e^{-x})-(e^x-e^{-x})(e^x+e^{-x})'}{(e^x+e^{-x})^2}$$

$$= \frac{(e^x+e^{-x})^2-(e^x-e^{-x})^2}{(e^x+e^{-x})^2}$$

$$= 1-\left(\frac{(e^x-e^{-x})^2}{(e^x+e^{-x})}\right)^2$$

$$= 1-\tanh^2(x)$$

$$= \operatorname{sech}^2(x)$$

<Derivative formula of hyperbolic function>

$\dfrac{d}{dx}\sinh(x)=\cosh(x)$ $\qquad$ $\dfrac{d}{dx}\cosh(x)=\sinh(x)$

$\dfrac{d}{dx}\tanh(x)=\operatorname{sech}^2(x)$ $\qquad$ $\dfrac{d}{dx}\operatorname{csch}(x)=-\operatorname{scsh}(x)\coth(x)$

$\dfrac{d}{dx}\operatorname{sech}(x)=-\operatorname{sech}(x)\tanh(x)$ $\qquad$ $\dfrac{d}{dx}\coth(x)=-\dfrac{1}{\sinh^2(x)}$

The hyperbolic functions in the sympy module are identical to the hyperbolic expressions above.

Ex) Differentiate follow functions?

1) $f(x)=2x^5\cosh(x)$

$f'(x)=10x^4\cosh(x)+2x^5\sinh(x)$

>>> f=2*x**5*cosh(x);f

2*x**5*cosh(x)

62

```
>>> diff(f, x)
2*x**5*sinh(x) + 10*x**4*cosh(x)
```

2) $g(x) = \dfrac{\sinh(x)}{x+1}$

$g'(x) = \dfrac{\cosh(x)(x+1) - \sinh(x)}{(x+1)^2}$

```
>>> g=sinh(x)/(x+1);g
sinh(x)/(x + 1)
>>> factor(diff(g, x))
(x*cosh(x) - sinh(x) + cosh(x))/(x + 1)**2
```

3) $1/x$

```
>>> diff(1/x, x)
-1/x**2
```

4) $(5x^3 - 7x + 1)^5$

```
>>> diff((5*x**3-7*x+1)**5, x)
(75*x**2 - 35)*(5*x**3 - 7*x + 1)**4
```

5) $\sin(3-6x)$

```
>>> diff(sin(3-6*x), x)
-6*cos(6*x - 3)
```

6) $\exp(x^2 - 9x)$

```
>>> diff(exp(x**2-9*x), x)
(2*x - 9)*exp(x**2 - 9*x)
```

1.3.6 Explicit and Implicit Function

An explicit function is a function form in which f(x) consists of independent variables in the form of y=f(x). The implicit function, on the other hand, is a form of function that contains both independent variables (x) and dependent variables (y), such as $x^2+y^2=a$.

The following function is an equation of a circle with a radius of three: Differentiate this expression against y.

From $x^2+y^2=9$, y'?

The above form is implicit function.

In order to convert the implicit function into an explicit function type, the remaining terms except for the dependent variable y is arranged by for y. Its explicit function is as follows.

$$y=\pm\sqrt{9-x^2}$$

The derivative of the explicit function must be executed separately for positive and negative cases.

```
>>> df1=diff(sqrt(9-x**2), x);df1
-x/sqrt(-x**2 + 9)
>>> df2=diff(-sqrt(9-x**2), x);df2
x/sqrt(-x**2 + 9)
```

The tangent equation at (2, $5^{0.5}$) is calculated by df1 of above results because y>0.

```
>>> slope=df1.subs(x, 2);slope
-2*sqrt(5)/5
```

Therefore, y=y₁+slope(x-x₁)

$$y=\sqrt{5}-\frac{2\sqrt{5}}{5}(x-2)$$

Another example, the form of the implicit function of the radius 5 function is as follows.

$x^2+y^2=25$

Of course, this function can be represented by a function as follows:

$$y=\pm\sqrt{25-x^2}$$

```
>>> x=symbols('x', real=True)
>>> y=Function('y')(x);y
y(x)
>>> eq=x**2+y**2-25;eq
x**2 + y(x)**2 - 25
>>> y1=solve(eq, y);y1
[-sqrt(-x**2 + 25), sqrt(-x**2 + 25)]
```

Differentiating the explicit functions are executed by separating positive and negative ranges.

$$y'=\begin{cases} \text{if } y>0, & \dfrac{d}{dx}\sqrt{25-x^2}=\dfrac{-x}{25-x^2} & (1) \\ \text{if } y\leq0 , & \dfrac{d}{dx}(-\sqrt{25-x^2})=\dfrac{x}{25-x^2} & (2) \end{cases}$$

```
>>> df1=diff(y1[0], x);df1 #=(2)
```

```
x/sqrt(-x**2 + 25)
>>> df2=diff(y1[1], x);df2 #=(1)
-x/sqrt(-x**2 + 25)
```

On the other hand, the derivative of the implicit function is as follows.

$$2x\frac{dx}{dx}+2y\frac{dx}{dy}=0$$

$$\rightarrow \quad \frac{dx}{dy}=-\frac{x}{y} \quad (3)$$

```
>>>deq=diff(eq, x);deq
2*x + 2*y(x)*Derivative(y(x), x)
>>> df=solve(deq, diff(y, x));df
[-x/y(x)]
```

The derivative of an implicit function can be applied to all y values without preconditions.

For example, for derivatives at position (3, 4)

For both functions, y>0, so it should be applied to (1).

On the other hand, the implicit function is applied to (3) without considering the preconditions of y.

```
>>> df2.subs(x, 3)
-3/4
>>> df[0].subs({x:3, y:4})
-3/4
```

\<code>

<u>Function()</u>: Sympy function to create function

The equation of the circle above is an implicit form. In this equation, the dependent variable y is a function with x as an independent variable, so we defined y by using the Function() function.

<u>solve()</u>: A sympy function to calculate the solution of an equation. The result is returned in list form.

To use an element in a list, call it with a list name [element index]. In python, indexes start at zero. Therefore, in the above case, the first solution is called with df [0].

Also, expressions passed as arguments to the solve() function can be represented in two formats.

$x^2+y^2-25=0$: Homogeneous equation

$x^2+y^2=25$: Non-homogeneous equation, Created by <u>Eq()</u>.

In general, solve() can use both types, but <u>diff()</u> requires a homogeneous solution. When using non-homogeneous, it returns only the expression, which is not evaluated as follows:

```
>>>eq=Eq(x**2+y**2,25);eq
Eq(x**2 + y(x)**2, 25)
>>> deq=diff(eq, x);deq
Derivative(Eq(x**2 + y(x)**2, 25), x)
```

In the above results, the following expressions are all the same.

Derivative(y(x), x)=y'(x)=diff(y, x)

Ex) Air is injected at 5 cm³/min in a ball shaped balloon. The rate of change of the radius in the balloon when the diameter of the balloon is 20 cm?

The volume of the sphere $=4/3\ \pi[r(t)]^3$

The radius at 20 cm diameter, $r(t)=20/2=10$

```
>>> t=symbols("t")
>>> r=Function('r')(t);r
r(t)
>>> v=Rational("4/3")*pi*r**3;v #volume=4/3·π·(r(t))³
4*pi*r(t)**3/3
>>> dv=diff(v, t);dv # (volume)'=4·π·(r(t))²(r(t))'
4*pi*r(t)**2*Derivative(r(t), t)
>>> root=solve(dv-5, diff(r, t));root #(r(t))'=5/(4·π·(r(t))²)
[5/(4*pi*r(t)**2)]
>>> drdt=root[0].subs(r, 10);drdt #(r(t))'=5/(4·π·10²)
1/(80*pi)
```

Ex) A 15 meter high ladder leans against the wall, and the distance from the wall to the ladder is 10 meters. In this case, if the ladder moves at 0.25 m/min toward the wall, how fast does the ladder move in units of wall height after 12 seconds?

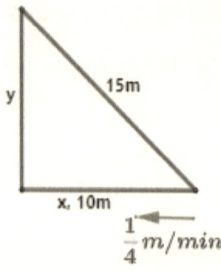

The formula for x and y by Pythagorean theorem is:

$x^2 + y^2 = 15^2$

$dx/dt=0.25 \rightarrow dx=12dt=12\cdot(1/4)=3$ in 12 sec

In 12 sec, x=10-3=7 m

Apply an implicit derivative to calculate dy for time t after 12 seconds.

$$2x\frac{dx}{dt}+2y\frac{dy}{dt}=0 \rightarrow \frac{dy}{dt}=-\frac{x}{y}\frac{dx}{dt}$$

```
>>> t=symbols("t")
>>> y=Function('y')(t);y
y(t)
>>> eq=x**2+y**2-15**2;eq
x(t)**2 + y(t)**2 - 225
```

In the above equation, the solution of y for x=7 is as follows.

```
>>> sol=solve(eq.subs(x, 7), y);sol
[-4*sqrt(11), 4*sqrt(11)]
```

y(length)>0 → y=sol[1]

dy/dt by differentiating eq is Derivative(y(t), t) in follows:

```
>>> deq=diff(eq, t);deqdt
2*x(t)*Derivative(x(t), t) + 2*y(t)*Derivative(y(t), t)
```

The conditions to be assigned in the above result are as follows.

x(t):7, dx/dt:-¼, y(t): sol[1]

```
>>> root=solve(deqdt, diff(y, t));root #dy/dt
[-x(t)*Derivative(x(t), t)/y(t)]
>>> dydt=root[0].subs({x:7, diff(x, t):-1/4, y:sol[1]})
>>> N(dydt, 4)
0.1319
```

dy/dt=0.1319 m/min

Ex) Two people are 50 m apart and one person moves at the speed of θ=0.01 radian/min. The rate of change of distance between two people when θ=0.5 radians?

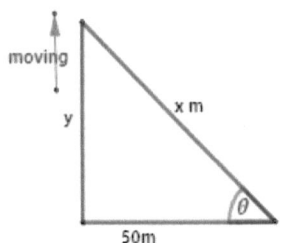

θ'=(dθ)/(dt)=0.01, 50/x=cos(θ(t)) → x' at x=50/cos(θ(t))?

x'=(dx)/(dt)=(50/cos(θ(t)))'

```
>>> t=symbols("t")
```

```
>>> r=Function("r")(t);r
r(t)
>>> f=50/cos(r);f
50/cos(r(t))
>>> df=diff(f, t);df
50*sin(r(t))*Derivative(r(t), t)/cos(r(t))**2
>>> N(d_f.subs({r:0.5, diff(r, t):0.01}), 4)
0.3113
```

Ex) In a conical water tank with a diameter of 5 m and a height of 14 m, water flows at a constant speed of 2 m³/hr.

1) When the water depth is 6 m, the rate of change of water height in the tank?

Water volume in tank V(t) is determined by radius(r(t)) and height(h(t)).

volume change rate V'(t)=2 m³/hr

$$V(t)=\frac{1}{3}\pi r^2(t)h(t)$$

$$V'(t)=\frac{2}{3}\pi r(t)\frac{dr(t)}{dt}h(t)+\frac{1}{3}\pi r^2(t)\frac{dh(t)}{dt}=2$$

From the problem, the ratio of radius to height can be calculated. That is,

r (t)/h(t) = 5/14.

```
>>> t=symbols("t")
>>> r=Function("r")(t);r
r(t)
>>> h=Function("h")(t);h
```

h(t)

```
>>> v=Rational('1/3')*pi*r**2*h;v
pi*h(t)*r(t)**2/3
>>> v2=v.subs(r,Rational("5/14")*h);v2
25*pi*h(t)**3/588
>>> dvdt=diff(v2, t); dvdt
25*pi*h(t)**2*Derivative(h(t), t)/196
>>> dhdt=solve(dvdt+2, diff(h, t));dhdt
[-392/(25*pi*h(t)**2)]
>>> dhdt[0].subs(h, 6)
-98/(225*pi)
>>> dhdt[0].subs(h, 6).evalf(4)
-0.1386
```

2) The rate of change of the radius of the water surface under the same conditions as above?

h(t)=(14/5)r(t) → r=(5·6)/14

```
>>> v3=v.subs(h,Rational("14/5")*r);v3
14*pi*r(t)**3/15
>>> dvdt=diff(v3, t); dvdt
14*pi*r(t)**2*Derivative(r(t), t)/5
>>> drdt=solve(dvdt+2, diff(r, t));drdt
[-5/(7*pi*r(t)**2)]
>>> drdt[0].subs(r, 5*6/14).evalf(4)
-0.04951
```

Ex) It is discharged at a constant speed of 6 m³/sec in an isosceles triangular water tank. Rate of change of height and width of water when the height of water is 1.2 m?

width(w)=5 m, height(h)=2 m, length(l)=8 m

$$w:h=5:2 \rightarrow w=\frac{5}{2}h$$

Water Volume: $V(t)=\frac{1}{2}w(t)\cdot h(t)\cdot l=4\frac{5}{2}h\cdot h=10h^2$

$$V'(t)=20h\frac{dh}{dt}$$

```
>>> t=symbols("t")
>>> w=Function("w")(t);w
w(t)
>>> h=Function("h")(t);h\f
h(t)
>>> v=4*w*h
```

1) h'(t)?

```
>>> v2=v.subs(w,Rational("5/2")*h);v2
10*h(t)**2
>>> dvdt=diff(v2, t); dvdt
20*h(t)*Derivative(h(t), t)
>>> dhdt=solve(dvdt-6, diff(h, t));dhdt
[3/(10*h(t))]
>>> dhdt[0].subs(h, 120).evalf(4)
0.002500
>>> dhdt[0].subs(h, 1.2).evalf(4)
0.2500
```

∴ h'(t)=0.25 m/sec

2) w'(t)?

   w=(5/2)·1.2 at h=1.2

>>> v2=v.subs(h,Rational("2/5")*w);v2
8*w(t)**2/5
>>> dvdt=diff(v2, t); dvdt
16*w(t)*Derivative(w(t), t)/5
>>> dwdt=solve(dvdt-6, diff(w, t));dwdt
[15/(8*w(t))]
>>> dwdt[0].subs(w, 5*1.2/2).evalf(4)
0.6250
∴ w(t)=0.625 m/sec

1.3.7 Higher Order Derivatives

Differentiate the following functions.

$f(x)=5x^3-3x^2+10x-5 \rightarrow f'(x)=15x^2-6x+10$

Here is another derivation of the above differentiated function.

$(f'(x))'=f''(x)=30x-6$

The first differentiated function f'(x) is called the first derivative, and the second derivative f''(x) is called the second derivative. Differentiating the second derivative results in a third derivative(f'''(x)).

$f'''(x)=30$

The fourth derivative is expressed differently.

$f^{(4)}(x)=(f'''(x))'=0$

If f(x) is an n-dimensional polynomial, it can be differentiated until the derivative of the function is zero.

$f^{(k)}(x)=0, \ k \geq n+1$

Derivatives higher than the second order are called high-order derivatives. The high-order derivative can be expressed as:

$$f'(x)=\frac{dy}{dx}, \quad f''(x)=\frac{d^2y}{dx^2}, \quad f'''(x)=\frac{d^3y}{dx^3}, \cdots$$

Higher-order derivatives can also be done using the sympy module function diff().

```
>>> x=symbols('x')
>>> f=5*x**3-3*x**2+10*x-5;f
5*x**3 - 3*x**2 + 10*x - 5
>>> diff(f, x)
15*x**2 - 6*x + 10
>>> diff(f, x, x)
6*(5*x - 1)
>>> diff(f, x, x, x)
30
>>> diff(f, x, 3)
30
```

Ex) Differentiate the following functions 4th ?

1) $f(x)=3x^2+8x^{0.5}+e^x$

```
>>> f=3*x**2+8*x**(1/2)+exp(x);f
8*x**0.5 + 3*x**2 + exp(x)
>>> diff(f, x, 4)
-7.5*x**(-3.5) + exp(x)
```

2) y=cos(x)

```
>>> diff(cos(x), x, 4)
cos(x)
```

3) $f(x)=\sin(3x)+e^{-2x}+\ln(7x)=\sin(3x)+e^{-2x}+\ln(7)+\ln(x)$

$$f'(x)=3\cos(3x)-2e^{-2x}+\frac{1}{x}$$

$$f''(x)=-9\sin(3x)+4e^{-2x}+\frac{1}{x^2}$$

$$f'''(x)=-27\cos(3x)-8e^{-2x}+\frac{2}{x^3}$$

$$f^{(4)}(x)=81\sin(3x)+16e^{-2x}+\frac{6}{x^4}$$

```
>>> f=sin(3*x)+exp(-2*x)+ln(7*x);f
log(7*x) + sin(3*x) + exp(-2*x)
>>> diff(f, x, 1)
3*cos(3*x) - 2*exp(-2*x) + 1/x
>>> diff(f, x, 2)
-9*sin(3*x) + 4*exp(-2*x) - 1/x**2
>>> diff(f, x, 3)
-27*cos(3*x) - 8*exp(-2*x) + 2/x**3
>>> diff(f, x, 4)
81*sin(3*x) + 16*exp(-2*x) - 6/x**4
```

Ex) y'' of $x^2 + y^4 = 10$?

$$2x + 4y^3 \frac{dy}{dx} = 0 \rightarrow \frac{dy}{dx} = -\frac{x}{2y^3}$$

$$\frac{d^2y}{dx^2} = -\frac{2y^3 - x \cdot 6y^2 \frac{dy}{dx}}{(2y^3)^2} = -\frac{3x^2 + 2y^4}{4y^7}$$

```
>>> y=Function("y")(x);y
y(x)
>>> eq=x**2+y**4-10;eq
x**2 + y(x)**4 - 10
>>> deq=diff(eq, x); deq
2*x + 4*y(x)**3*Derivative(y(x), x)
>>> dy=solve(deq, diff(y, x));dy
[-x/(2*y(x)**3)]
>>> dy2=diff(dy[0], x);dy2
3*x*Derivative(y(x), x)/(2*y(x)**4) - 1/(2*y(x)**3)
>>> factor(dy2.subs(diff(y, x), dy[0]))
-(3*x**2 + 2*y(x)**4)/(4*y(x)**7)
```

1.4 Application of differential

1.4.1 Rate of change

Ex) Any point in the following function that has no change ?

g(x)=5-6x-10cos(2x)

No change means a point where the derivative result is zero.

```
>>> x=symbols("x", real=True)
>>> g=5-6*x-10*cos(2*x);g
-6*x - 10*cos(2*x) + 5
>>> dg=diff(g, x); dg
20*sin(2*x) - 6
>>> sol=solve(dg, x); sol
[-asin(3/10)/2 + pi/2, asin(3/10)/2]
>>> [N(i, 4) for i in sol]
[1.418, 0.1523]
```

<code>

In the above code, the result of solve() is returned as the value of asin(). This value can be converted to a real number using the N() or .evalf() function.

Also, if there are several solutions, repeat the same command to convert the values to real numbers. In this case, the loop statement can be used. The list comprehension syntax, which is a brief form of the loop statement, is used.

Because the function in this example includes cos(2x), it repeats a certain interval periodically depending on the characteristics of the trigonometric function. Therefore, as shown in Figure 1.7, the point where the rate of change is zero is very much in addition to the above results.

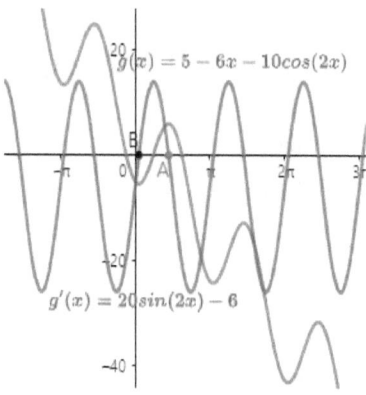

Figure 1.7. g(x)=5-6x-10cos(2x) and g'(x).

The cos(x) function has a period of 2π, but for cos(2x), the period is π (Figure 1.7).

2x=2π → x=π

Therefore, all points where the derivative of the above equation becomes 0 can be expressed as follows.

1.428 ± nπ or 0.1523 ± nπ, n=0, ±1, ±2, ...

Ex) Describe the increment and decrement intervals of the following functions.

79

$$A(t) = 27t^5 - 45t^4 - 130t^3 + 150$$

```
>>> t=symbols('t')
>>> A=27*t**5-45*t**4-130*t**3+150;A
27*t**5 - 45*t**4 - 130*t**3 + 150
>>> dA=diff(A, t); dA
135*t**4 - 180*t**3 - 390*t**2
>>> sol=solve(dA, t); sol
[0, 2/3 + sqrt(30)/3, -sqrt(30)/3 + 2/3]
```

The above result shows the point where the derivative is zero, that is, the point where the rate of change is zero. The sign change of the derivative function based on these points is a clue to judge the increase and decrease of the function.

```
>>> rng=np.append(sol, np.arange(np.min(sol)-1, np.max(sol)+1,
1));rng
array([0, 2/3 + sqrt(30)/3, -sqrt(30)/3 + 2/3, -sqrt(30)/3 - 1/3, -
sqrt(30)/3 + 2/3, -sqrt(30)/3 + 5/3, -sqrt(30)/3 + 8/3, -sqrt(30)/3
+ 11/3, -sqrt(30)/3 + 14/3], dtype=object)
>>> re={}
>>> re1={}
>>> for i in np.sort(rng):
...    re[i]=dA.subs(t, i).evalf(4)
...    if re[i].is_integer or re[i].is_real:
...        if re[i]<=10**(-10) and re[i]>=-10**(-10):
...            re1[i]=0
...        elif re[i]>0:
...            re1[i]="+"
```

```
...        elif re[i]<0:
...            re1[i]="-"
...        else:
...            re[i]=0
...      else:
...        re1[i]=re[i]
>>> pd.DataFrame([re1.keys(), re1.values()]).T
                0              1
0  -sqrt(30)/3 - 1/3      +
1  -sqrt(30)/3 + 2/3      0
2  -sqrt(30)/3 + 5/3      -
3          0             0
4  -sqrt(30)/3 + 8/3      -
5 -sqrt(30)/3 + 11/3      -
6   2/3 + sqrt(30)/3      0
7 -sqrt(30)/3 + 14/3      +
```

Decrease interval;

-sqrt(30)/3+2/3<t<2/3+sqrt(30)/3

Increase interval;

-sqrt(30)/3+2/3>t and 2/3+sqrt(30)/3<t

<code>

Some values must be assigned to the derivative function to determine the change in sign based on the solution(s) of the expression. This process was performed using a loop. The scope for using the loop was created by functions of the four numpy modules.

np.append(x, y): Create an object by adding y after the object x.

np.arange(start, end, step): creates a sequence

np.min(x): Return minimum.

np.max(x): Return maximum.

np.sort(x): Sort objects x in ascending order.

The results of the loop are stored in two objects 're' and 're1' in dictionary format.

re: save values

re1: Save the sign

For a computer operation, 2/3+sqrt (30)/3 represents a nonzero value when assigned to a derivative function. However, this value is a solution of the expression and should be treated as zero. This difference is due to the precision of handling the float, and the result is forced to zero using the 'if conditional'. It is a command to judge 0 as 10^{-10} or less.

```
>>> dA.subs(t, 2/3+sqrt(30)/3).evalf(4)
-1.133e-13
```

DataFrame(): Converts an object to a pandas object (a table type with rows and columns) as a function of the pandas module.

Write the above code as a function to apply to other problems.

```python
def FunIncDecS(fun, var, criticp, rev=0.5, n=3):
    rng=np.append(criticp, np.arange(np.min(criticp)-rev,
np.max(criticp)+0.1+rev, rev))
    re={}
    re1={}
    for i in np.sort(rng):
        i1=N(i, n)
        re[i1]=fun.subs(var, i).evalf(n)
        if re[i1].is_integer or re[i1].is_real:
            if re[i1]<=10**(-10) and re[i1]>=-10**(-10):
                re1[i1]=0
            elif re[i1]>0:
                re1[i1]="+"
            elif re[i1]<0:
                re1[i1]="-"
            else:
                re[i1]=0
        else:
            re1[i1]=re[i1]
    return([pd.DataFrame([re1.keys(), re1.values()]).T,
    pd.DataFrame([re.keys(), re.values()]).T])
```

Arguments;

fun: Function to apply

var: the variable to pass the value to

critip: the solution of the derivative function

ref: Range

n: the number of significant digits in the returned value.

Ex) Cars A and B are located 500 m apart in the west and east, respectively. A moves at a speed of 35 m/h toward B and B moves at a speed of 50 m/h to the south. Rate of change of distance between two cars after 3 hours? Increase or decrease?

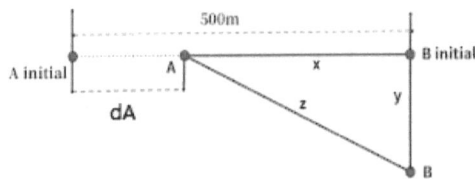

The above problem is to calculate the rate of change of z in above figure. The above x, y, and z can be recognized as a function of t.

$x^2(t)+y^2(t)=z^2(t)$

$(x(t))'=-35$, $(y(t))=50$

After 3 hour, x=500-35·3=395, y=50·3=150

```
>>> t=symbols('t')
>>> x=Function('x')(t);x
x(t)
>>> y=Function('y')(t);y
y(t)
>>> z=Function('z')(t);z
```

z(t)

```
>>> eq=x**2+y**2-z**2;eq
x(t)**2 + y(t)**2 - z(t)**2
>>> deq=diff(eq, t);deq
2*x(t)*Derivative(x(t), t) + 2*y(t)*Derivative(y(t), t) -
2*z(t)*Derivative(z(t), t)
>>> dz=solve(deq, diff(z, t)); dz
[(x(t)*Derivative(x(t), t) + y(t)*Derivative(y(t), t))/z(t)]
>>> x1=500-35*3
>>> y1=50*3
>>> z1=(x1**2+y1**2)**0.5;z1
422.5221887664599
>>> dz[0].subs({x:x1, diff(x, t):-35, y:y1, diff(y, t):50, z:z1})
-14.9696280293957
```

1.4.2 Critical point

<Definition>

If the derivative of the function f(x) for any point c is 0, then the

point c is called the critical point.

  $f'(c)=0 \to c(s)$ is the critical point(s).

Ex) Critical point(s) of the followings ?

1) $f(x)=6x^5+33x^4-30x^3 +100$

   $f'(x) =30x^4+132x^3-90x^2$

```
>>> x=symbols('x')
>>> f=6*x**5+33*x**4-30*x**3+100;f
6*x**5 + 33*x**4 - 30*x**3 + 100
>>> df=diff(f, x);df
30*x**4 + 132*x**3 - 90*x**2
>>> cp=solve(df, x);cp
[-5, 0, 3/5]
```

2) $g(t)= t^{2/3}(2t-1)$

$$g'(t)=\frac{2}{3}t^{-\frac{1}{3}}(2t-1)+2t^{\frac{2}{3}}=\frac{10t-2}{3t^{\frac{1}{3}}}$$

```
>>> t=symbols('t')
>>> g=t**(Rational("2/3"))*(2*t-1);g
t**(2/3)*(2*t - 1)
>>> dg=diff(g, t);dg
2*t**(2/3) + 2*(2*t - 1)/(3*t**(1/3))
>>> factor(dg)
2*(5*t - 1)/(3*t**(1/3))
```

The above solution uses the solve() function. If t=0 in this equation, the denominator is zero and is undefined. Therefore, the solution is calculated under the condition $t \neq 0$.

```
>>> cp=solve(dg, t);cp
[1/5]
```

3) $R(x)=\dfrac{x^2+1}{x^2-x-6}$

$R'(x)=\dfrac{2x(x^2-x-6)-(x^2+1)(2x-1)}{(x^2-x-6)^2}$

```
>>> x=symbols('x')
>>> r=(x**2+1)/(x**2-x-6);r
(x**2 + 1)/(x**2 - x - 6)
>>> dr=diff(r, x); dr
2*x/(x**2 - x - 6) + (-2*x + 1)*(x**2 + 1)/(x**2 - x - 6)**2
>>> cp=solve(dr, x);cp
[-7 + 5*sqrt(2), -5*sqrt(2) - 7]
```

4) $y=6x-4\cos(3x)$

   $y'=6+12\sin(3x)$

```
>>> x=symbols('x')
>>> y=6*x-4*cos(3*x);y
6*x - 4*cos(3*x)
>>> dy=diff(y, x);dy
12*sin(3*x) + 6
>>> cp=solve(dy, x);cp
[-pi/18, 7*pi/18]
>>> [N(cp[0], 4), N(cp[1], 4)]
[-0.1745, 1.222]
```

As shown in Figure 1.8, the above function has a $2\pi/3$ period due to cos(3x). Therefore, if a specific range is not specified, the point where the derivative is zero is as follows.

[-0.1745+2nπ/3, 1.222+2nπ/3], n=0, ±1, ±2,...

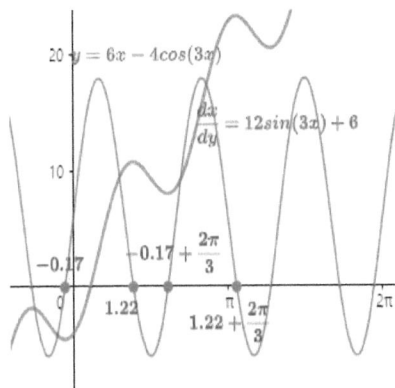

Figure 1.8. y=6x-4cos(3x) and y'.

5) $h(t) = 10t \cdot \exp(3-t^2)$

$h'(t) = 10\exp(3-t^2) + 10t(-2t)\exp(3-t^2)$

```
>>> t=symbols('t')
>>> h=10*t*exp(3-t**2);h
10*t*exp(-t**2 + 3)
>>> dh=diff(h, t);dh
-20*t**2*exp(-t**2 + 3) + 10*exp(-t**2 + 3)
>>> cp=solve(dh, t);cp
[-sqrt(2)/2, sqrt(2)/2]
```

6) $f(x) = x^2\ln(3x) + 6$

$f'(x) = 2x \cdot \ln(3x) + x$

```
>>> x=symbols('x')
>>> f=x**2*log(3*x)+6;f
x**2*log(3*x) + 6
>>> df=diff(f, x);df
```

88

2*x*log(3*x) + x

>>> cp=solve(df, x);cp

[exp(-1/2)/3]

7) $f(x)=x \cdot \exp(x^2)$

   $f'(x)=\exp(x^2)(2x^2+1)$

>>> x=symbols('x', real=True)

>>> f=x*exp(x**2);f

x*exp(x**2)

>>> df=diff(f, x);df

2*x**2*exp(x**2) + exp(x**2)

>>> factor(df)

(2*x**2 + 1)*exp(x**2)

df(f'(x))≠0 → Critical point of f(x) doesn't exist.

The solve() function returns zoo as follows: In sympy, zoo represents infinity. This means that there is no solution in the real scope.

>>> cp=solve((2*x**2 + 1)*exp(x**2), x);cp

[zoo]

1.4.3 Maximum and Minimum

The maximum and minimum values of a function are defined as follows.

<Definition>

1) If $f(x) \leq f(c)$ holds for all x of the set interval, it is said to have the *(absolute) maximum* of $f(x)$ at $x = c$.

2) If $f(x) \leq f(c)$ holds for any open interval x, it is said to have a *local maximum* at $x = c$.

3) If $f(x) \geq f(c)$ holds for all x in the set interval, it is said to have the *(absolute) minimum* of $f(x)$ at $x = c$.

4) If $f(x) \geq f(c)$ holds for any open interval x, it is said to have a *local minimum* at $x = c$.

The interval between a and b is expressed as follows.

$(a, b) \rightarrow a < x < b$: open interval

$[a, b] \rightarrow a \leq x \leq b$: closed interval

These values are called extrema. Note that the local maxima or minima can not be determined in the lower and upper limits of the closed interval, since the local extrema are values determined in the open interval.

Ex) Extrema of the followings?

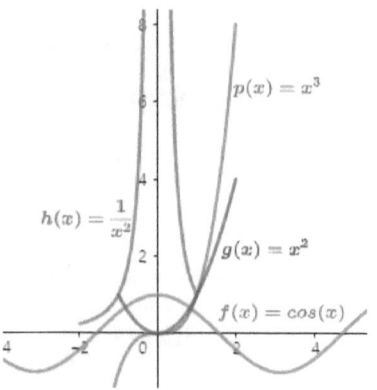

Figure 1.9. The graphs of followings.

From Figure 1.9,

1) $g(x)=x^2$, $[-1, 2]$

Maximun: f(2)

However, since this interval is the closed interval and the point (2) is the upper limit of the interval, it can not be the local maximum value.

Minimum: f(0)

The value is local minimum because the point is between -1 and 2.

2) $p(x)=x^3$ at all x, and $[-2, 2]$

Because $-\infty < p(x) < \infty$, the extrema can not be determined in all x. Since $[-2, 2]$ is a closed interval, g(2) and g(-2) are the maximum and minimum values, respectively, but not the local extrema.

3) $f(x)=\cos(x)$

The function f(x) repeats between 1 and -1 with a period of 2π.

Therefore, there are several maximum and minimum values.

Maximum, local maximum = ..., $-4\pi, -2\pi, 0, 2\pi, 4\pi, ...$

Minimum, local minimum = ..., $-3\pi, -\pi, 0, \pi, 3\pi, ...$

4) $h(x)=x^{-2}$ [-2, 1]

h(x) is discontinuous at x=0. In other words, it increases to infinity. Therefore, the maximum value of this interval and the local maximum value do not exist. However, there is a minimum at x=-2. However, since the point is not included in the opening interval, it can not be the local minimum value.

The relationship between the minimum value of the function and the derivative is summarized as follows.

<Theorem>

If f(x) has a local minimum at x=c and f'(c) exists, then x=c is the critical point of f(x).

1) if $h<0$, $f(c) \geq f(c+h) \rightarrow f(c+h)-f(c) \leq 0$

$$\lim_{h \to 0^+} \frac{f(c+h)-f(c)}{h} \leq \lim_{h \to 0^+} 0 = 0 \rightarrow f'(c) \leq 0$$

2) if $h>0$, $f(c) \leq f(c+h) \rightarrow f(c+h)-f(c) \geq 0$

$$\lim_{h \to 0^-} \frac{f(c+h)-f(c)}{h} \geq \lim_{h \to 0^-} 0 = 0 \rightarrow f'(c) \geq 0$$

$\therefore 0 \leq f'(c) \leq 0 \rightarrow f'(c)=0$

For $g(x)=x^2$, the local minimum is zero, as shown in Figure 1.9.

At x=0, g(x) is continuous and can be differentiated. Therefore, the local minimum point(x = 0) is the critical point.

However, for h(x)=x^{-2}, there is no relative minimum. Therefore, the critical point also does not exist.

The extrema of f(x) in [a, b] can be found in the following procedure

1) Test that the function is continuous in the interval.

2) Find all critical points in interval [a, b].

3) Compares the function value(s) at the critical point with the function values corresponding to the upper and lower limits of the specified interval.

Ex) The extrema of g(x)=2x^3+3x^2-12x+4 at [-1, 3] ?

```
>>> x=symbols('x')
>>> g=2*x**3+3*x**2-12*x+4;g
2*x**3 + 3*x**2 - 12*x + 4
>>> dg=diff(g, x);dg
6*x**2 + 6*x - 12
>>> cp=solve(dg, x);cp #critical point
[-2, 1]
>>> g_cp=[g.subs(x, i) for i in cp];g_cp
[24, -3]
```

The critical point in the specified interval is 1, and g(x) corresponding to that point is -3. The values corresponding to the upper and lower limits of the interval are as follows.

```
>>> g_ep=[g.subs(x, i) for i in [-1, 3]];g_ep
[17, 49]
```

Here is a comparison of the results returned from the above process:

g(1)<g(-1)<g(3)

g(3) is the maximum value because it corresponds to the upper limit of the specified interval, but g(1) is the local minimum value corresponding to the x value in the interval.

　Maximum value: g(3)=49 at upper limit point 3

　Local minimum: g(1)=- 3 at critical point 1

Ex) The population increase or decrease with time t(month) in one area follows the following function.

P(t)=3t+sin(4t)+100

Maximum population and minimum population for four months?

```
>>> t=symbols('t')
>>> p=3*t+sin(4*t)+100;p
3*t + sin(4*t) + 100
>>> ep=[0, 4] #interval
>>> dp=diff(p, t);dp
```

4*cos(4*t) + 3

>>> cp=solve(dp, t);cp #critical point

[-acos(-3/4)/4 + pi/2, acos(-3/4)/4]

Due to sin(4t) in p(t), the period of the function is π/2.

4t=2π → t=π/2

Therefore, the critical points of the function can be represented by cp±nπ/2(n=0, 1, 2, 3, ...) and should be within interval [0, 4].

>>> cp1=np.array([[N(i+j*pi/2, 3) for i in cp for j in range(5)]]);cp1
array([[0.966, 2.54, 4.11, 5.68, 7.25, 0.605, 2.18, 3.75, 5.32, 6.89],
dtype=object)
>>> cp2=cp1[cp1<=4]
>>> cp2=cp2[cp2>=0];cp2
array([[0.966, 2.54, 0.605, 2.18, 3.75], dtype=object))
>>> p_cp=[N(p.subs(t, i), 3) for i in cp2];p_cp
[102., 107., 102., 107., 112.]

The following are the p(t) values corresponding to the upper and lower bounds of the specified interval.

>>> p_ep=[N(p.subs(t, i),3) for i in ep];p_ep
[100., 112.]

Comparing the p_cp and p_ep values above, the maximum population is 112 and the minimum population is 100. In other words,

Maximum and local maximum: 112

Local minimum: 100

95

<code>

The elements of the numpy object can be called using a comparison statement. That is, cp1<=4 returns 'True' if each element is bound to a condition, or 'False' otherwise.

```
>>> cp1<=4
array([ True,  True, False, False, False,  True,  True,  True,
False,False])
```

Applying the above result as an index returns only the 'True' part.

```
>>> cp1[cp1<=4]
array([[0.966, 2.54, 0.605, 2.18, 3.75], dtype=object)
```

Ex) Assume that the annual(t) amount of a bank account follows the function A(t).

$$A(t)=2000-10t\cdot\exp(5-t^2/8)$$

Maximum and minimum amount for 10 years after opening an account?

```
>>> t=symbols('t')
>>> A=2000-10*t*exp(5-t**2/8);A
-10*t*exp(-t**2/8 + 5) + 2000
>>> interval=[0, 10]
>>> dA=diff(A, t);dA
5*t**2*exp(-t**2/8 + 5)/2 - 10*exp(-t**2/8 + 5)
>>> cp=solve(dA, t);cp
```

[-2, 2]

time t>0 → cp=2

```
>>> A_cp=A.subs(t, 2).evalf(4);A_cp
199.7
```

'A' in the lower and upper bounds of the interval is as follows.

```
>>> A_interval=[A.subs(t, i).evalf(5) for i in interval];A_interval
[2000.0, 1999.9]
```

As a result of the calculation above, the minimum value is 199.7, which corresponds to critical point 2. In other words, the minimum value is 2 years after opening the account, and the maximum value is 2000 at the time of initial opening.

1.4.4 Graph shape

The maximum and minimum values of the specified interval(I) are calculated from the derivative of the function. In addition, the overall shape of the graph can be estimated from the change.

At interval I,
$x_1 < x_2$ → $f(x_1) < f(x_2)$: increase ↔ $f'(x) > 0$
$x_1 < x_2$ → $f(x_1) > f(x_2)$: decrease ↔ $f'(x) < 0$
$x_1 = x_2$ → $f(x_1) = f(x_2)$: critical point ↔ $f'(x) = 0$

If $x_1 < x_2$ at two points x_1 and x_2 belonging to interval I, there is a c value between them (see intermediate value theorem). That is, $x_1 < c < x_2$. In this condition, c satisfying $x_1 + c \to x_2$ can be set. This relationship can be expressed as:

$$\lim_{c \to 0} \frac{f(x_1+c)-f(x_1)}{(x_1+c)-x_1} = f'(c)$$

$$f'(c) = \frac{f(x_2)-f(x_1)}{x_2-x_1} \;\to\; f(x_2)-f(x_1)=f'(c)(x_2-x_1)$$

$$\Rightarrow \begin{cases} \text{if } f'(c)>0, \; f(x_2)>f(x_1) \\ \text{if } f'(c)<0, \; f(x_2)<f(x_1) \\ \text{if } f'(c)=0, \; f(x_2)=f(x_1) \end{cases}$$

Ex) Increment and decrement intervals of the following functions?

$$f(x) = -x^5 + (5/2)x^4 + (40/3)x^3 + 5$$

```
>>> x=symbols('x')
>>> f=-x**5+Rational("5/2")*x**4+Rational("40/3")*x**3+5 >>> f
-x**5 + 5*x**4/2 + 40*x**3/3 + 5
>>> df=diff(f, x);df
-5*x**4 + 10*x**3 + 40*x**2
>>> cp=solve(df, x); cp
[-2, 0, 4]
```

<code>

Rational(): Keeps the form of fraction x

$f'(c)=0$ at $c = -2, 0, 4$

By examining the sign of the derivative function f '(x) around this value, the increase or decrease of the function is determined. Apply the previously created user-defined function FunIncDecS() for this decision.

```
>>> re=FunIncDecS(df, x, cp)[0]; re
        0   1
0  -2.5    -
1    -2    0
2  -1.5    +
3  -0.5    +
4     0    0
5   0.5    +
6   1.5    +
7   2.5    +
8   3.5    +
9     4    0
10  4.5    -
```

The following 'fval' is the coordinate value determined by the function value corresponding to each value of the above result 're'. Based on these coordinates, Figure 1.10 can be estimated.

```
>>> fval=[(round(i, 2), round(f.subs(x, i), 2)) for i in re.iloc[:,0]];fval
[(-2.5, -8.02), (-2.0, -29.67), (-1.5, -19.75), (-0.5, 3.52), (0.0, 5.0), (0.5,
6.79), (1.5, 55.06), (2.5, 213.33), (3.5, 426.6), (4.0, 474.33), (4.5,
399.88)]
```

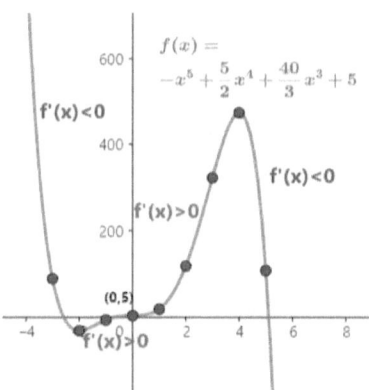

Figure 1.10. -x⁵+(5/2)x⁴+(40/3)x³+5.

The following relationships are summarized from the above process.

If the point c is the critical point of f (x), that is, f'(c) = 0
1) f'(x)>0 at x<c, f'(x)<0 at x>c → f(c) is local maximum.
2) f'(x)<0 at x<c, f'(x)>0 at x>c → f(c) is local minimum.

Ex) Graph shape of following function?

$$g(x)=\frac{x(x^2-4)}{3}$$

```
>>> x=symbols('x')
>>> g=x*(x**2-4)**1/3;g
x*(x**2 - 4)/3
>>> dg=diff(g, x);dg
x**2 - 4/3
>>> cp=solve(dg, x); cp #critical point
```

[-2*sqrt(3)/3, 2*sqrt(3)/3]
>>> re=FunIncDecS(dg, x, cp)[0]; re

```
     0   1
0  -1.65  +
1  -1.15  0
2  -0.655 -
3   0.345 -
4   1.15  0
5   1.35  +
```

Figure 1.11 can be estimated by plotting the values of g (x) corresponding to each section of the result above.

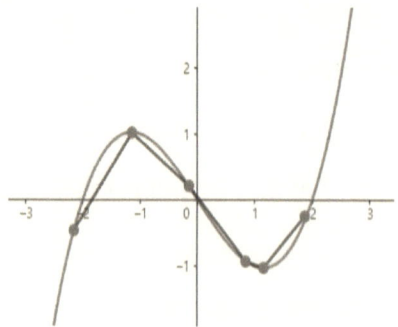

Figure 1.11. x(x²-4)/3.

In general, concavity graphs can be classified into four types as shown in Figure 1.12.

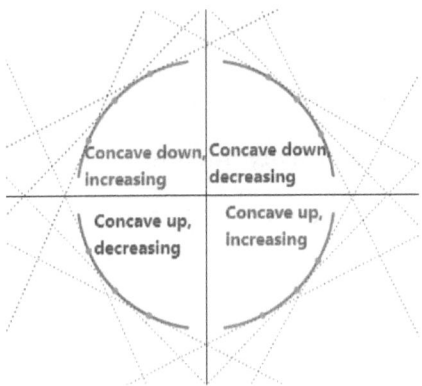

Figure 1.12. Types of concavity graphs.

The type of concave graph shown in Figure 1.12 is defined as follows.

In the specified interval,

1) If the tangent line is located at the bottom of the function graph, it is concave up type.

2) When the tangent line is located at the top of the function graph, it is concave down type.

3) If the function is continuous at point c and the concavity of the graph changes at that point, the point c is called the inflection point.

The concave form of the graph defined above can be related to the second derivative and can be expressed as:

In the specified interval,

102

(1) If f''(x)> 0 for all x, f(x) is concave up in that interval.

(2) If f''(x) <0 for all x, then f(x) in that interval is concave down.

(3) The change in concavity means a change in the second derivative function as defined in 1) and 2). That is, the point at which the sign of f''(x) changes is called the inflection point.

<Proof of definition 1)>

If the graph shape is concave up, the tangent line is located below f(x).

Tangent line at the point a: $y=f(a)+f'(a)(x-a)$

Therefore, the function value f (x) at an arbitrary point x is larger than any value generated by the tangent equation.

$f(x)>f(a)+f'(a)(x-a)$

According to the <u>intermediate theory</u>, the slope of a tangent to a point c between a and x equals the slope of the line between a and x.

$$a<c<x \rightarrow f'(c)=\frac{f(x)-f(a)}{x-a}$$

$f(x)=f(a)+f'(c)(x-a)$ ①

If f''(x)>0, f'(x)>0. That is, f '(x) is increasing in that interval.

Therefore, if a<c and f'(x)>0,

$f'(a)<f'(c)$ ②

x-a>0, so the following is true.

②×(x-a)+f(a) → $f(a)+f'(a)(x-a) <f(a)+f'(c)(x-a)=f(x)$ ③

$\therefore$ f"(x)>0 $\rightarrow$ f(a)+f'(a)(x-a) < f(x)

The tangent values are present under the function f(x). That is, concave up.

Ex) Inflection point, increase and decrease interval of function f(x)?

$h(x) = 3x^5 - 5x^3 + 3$

```
>>> x=symbols("x")
>>> h=3*x**5-5*x**3+3;h
 3*x**5 - 5*x**3 + 3
>>> dh=diff(h, x);dh
 15*x**4 - 15*x**2
>>> ddh=diff(h, x, 2);ddh #=diff(dh, x)
 30*x*(2*x**2 - 1)
```

1) The critical point of the function h(x), and the sign of the first-order derivative function of that function

```
>>> cp=solve(dh, x);cp
[-1, 0, 1]
>>> dhsig=FunIncDecS(dh, x, cp)[0];dhsig
```

		0	1
0	-1.50	+	
1	-1.00		0
2	-0.500	-	
3	0		0
4	0.500	-	

| 5 | 1.00 | 0 |
| 6 | 1.50 | + |

Increase: $-\infty<x<-1$, $1<x<\infty$

Decrease: $-1<x<0$, $0<x<1$

From the result, the local maximum and minimum of h(x) are h(-1) and h(1), respectively. The value of h(x) from each point of 'dhisg' above is calculated as follows.

```
>>> hval=FunIncDecS(h, x, cp)[1];hval
          0      1
0  -1.5000  -2.9063
1  -1.0000   5.0000
2  -0.50000  3.5313
3   0         3.0000
4   0.50000   2.4688
5   1.0000    1.0000
6   1.5000    8.9063
```

2) The inflection point is the point where the sign of the second derivative function is switched. Therefore, the sign of h''(x) can be checked around critical point(s).

```
>>> [N(ddh.subs(x, i), 3) for i in [-1.1, -1, -0.9, -0.1, 0, 0.9, 1, 1,1]]
[-46.9, -30.0, -16.7, 2.94, 0, 16.7, 30.0, 30.0, 30.0]
```

The result shows that the sign of h"(x) changes around zero. Therefore, the critical point 0 can be called the inflection point, and the sign of h''(x) at the other critical point -1, 1 does not change. Here is the summary.

Around -1, h''(x)<0: concave down ⇒ h'(x): "+" →"-" ⇒ local maximum

Around 0, h''(x)≅0: inflection point(sign change) ⇒ h'(x) : No sigh change ⇒ can define neither local maximum nor minimum

Around 1, h''(x)>0 :concave up ⇒ h'(x) : "+" →"-" ⇒ local minimum

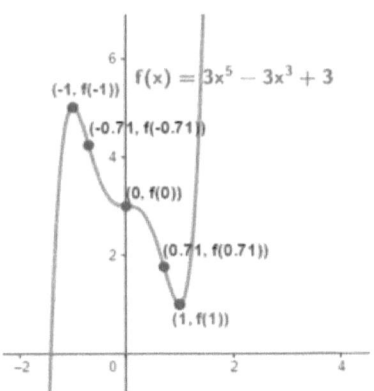

Figure 1.13. $3x^5-5x^3+3$.

Figure 1.13, which is the graph of h(x), shows the above relationship.

From the above result, if f"(c) is possible at the critical point c of the function f(x), we can see the following as the second derivative around each critical point.

1) if f"(points around c)<0, no inflection point exists → f(c) local minimum value

2) if f"(c around points)>0, no inflection point → f(c) local maximum value

3) If the sign of the values of f"(the points around c) is changed, there is an inflection point → Local minimum and maximum values can not be defined

Ex) The graph shape of f(x)=x⁴?

```
>>> x=symbols("x")
>>> f=x**4;f
 x**4
>>> df=diff(f, x);df
 4*x**3
>>> cp=solve(df, x);cp
 [0]
>>> re=FunIncDecS(df, x, cp)[0]; re
     0 1
0 -0.500 -
1   0 0
2 0.500 +
```

Based on critical point, f'(x) decreases and then increase. At this point, a local minimum value is formed. If the critical point is local minimum, the graph should be in the form of concave up in that interval. That is, f"(x)> 0.

```
>>> ddf=diff(f, x, 2);ddf
12*x**2
>>> [N(ddf.subs(x, i), 3) for i in [-0.1, 0, 0.1]]
[0.120, 0, 0.120]
```

Ex) $f(x) = -x^4$

```
>>> f=-x**4;f
-x**4
>>> df=diff(f, x);df
-4*x**3
>>> cp=solve(df, x);cp
[0]
>>> re=FunIncDecS(df, x, cp)[0]; re
       0 1
0 -0.500 +
1     0 0
2 0.500 -
>>> ddf=diff(f, x, 2);ddf
-12*x**2
>>> [N(ddf.subs(x, i), 3) for i in [-0.1, 0, 0.1]]
[-0.120, 0, -0.120]
```

No inflection point

Based on f'(0)=0, the sign of f'(x) changes '+' to '-'. Also, since the secondary differential around the critical point is f"(x)<0, it is concave down in that interval. That is, f(0) is local maximum.

Ex) $f(x) = -x^3$

```
>>> f=-x**3;f
-x**3
>>> df=diff(f, x);df
-3*x**2
>>> cp=solve(df, x);cp
[0]
>>> re=FunIncDecS(df, x, cp)[0]; re
     0 1
0 -0.500 -
1    0 0
2  0.500 -
>>> ddf=diff(f, x, 2);ddf
-6*x
>>> [N(ddf.subs(x, i), 3) for i in [-0.1, 0, 0.1]]
[0.600, 0, -0.600]
```

There is no sign change of f'(x) around the critical point. Therefore, both local minimum and maximum values can not be defined. Also, the sign of f"(x) is switched around the critical point. In other words, the critical point becomes an inflection point.

Consequently, the local maximum or minimum value of f(x) in that interval can not be defined. This is equivalent to the result from f'(x).

The graph for the above three functions is shown in Figure 1.14.

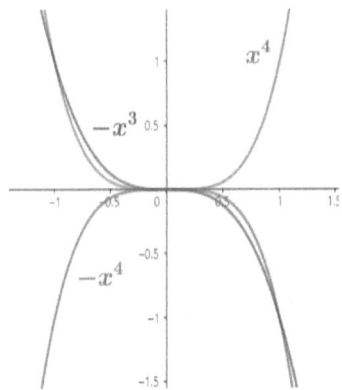

Figure 1.14. Graphs of x^4, x^{-4}, and x^{-3}.

Ex) Is the local minimum or maximum value at the critical point of the next function?

$f(t) = t(6-t)^{2/3}$

```
>>> t=symbols("t")
>>> f=-t*(6-t)**Rational("2/3");f
-t*(-t + 6)**(2/3)
>>> df=diff(f, t);df
2*t/(3*(-t + 6)**(1/3)) - (-t + 6)**(2/3)
>>> factor(df)
(5*t - 18)/(3*(-t + 6)**(1/3))
```

110

```
>>> cp=solve(df, t);cp
 [18/5]
>>> re=FunIncDecS(df, t, cp)[0]; re
    0 1
0 3.10 -
1 3.60 0
2 4.10 +
>>> ddf=diff(f, t, 2);factor(ddf)
-2*(5*t - 36)/(9*(-t + 6)**(4/3))
>>>[N(ddf.subs(t, i), 3) for i in [19/5, 18/5, 20/5]]
 [1.32, 1.24, 1.41]
```

Based on critical point 3.60(18/5), it shows "decrease→0→increase" form. Also, since f''(x)>0 based on that point, f(x) is a concave up form. Therefore f(18/5) is the local minimum value.

Ex) The fence will be installed around the house as shown in the following figure. Determine the area that can be installed the largest with 500 m of fence material?

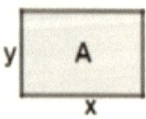

Area for installing fence: A=xy

x+2y=500→x=500-2y

111

A(y)=(500-2y)y

The maximum area is determined by the maximum value of the above function.

```
>>> y=symbols("y")
>>> A=(500-2*y)*y;A
y*(-2*y + 500)
>>> dA=diff(A, y);dA
-4*y + 500
>>> cp=solve(dA, y);cp
[125]
>>> FunIncDecS(dA, y, cp)[0]
          0 1
0 124.500 +
1    125 0
2 125.500 -
>>> ddA=diff(A, y, 2);ddA
-4
```

Based on critical point, the sign of A'(y) changes '+' to '-'. Since A"(y)<0, the graph of A(y) is concave down type, so the maximum value is determined at y = 125.

```
>>> A.subs(y, 125)
31250 m²
```

Ex) A rectangular box with a length of three times the width and a volume of 50 cubic meters is produced. The cost of making the side is 60\$/m², and the cost of making the top and bottom is 100\$/m². Determine the length(l), width(w), and height(h) of the box for the minimum cost of this box?

The total cost of box making: C=100(2wl)+60(2hw+2lh)

Condition: l=3w, 50=lwh

```
>>> l,w,h=symbols("l,w,h", real=True)
>>> l=3*w;l
 3*w
>>> h=50/(l*w);h
 50/(3*w**2)
>>> C=100*(2*w*l)+60*(2*h*w+2*l*h);C
 600*w**2 + 8000/w
>>> dC=diff(C, w);dC
 1200*w - 8000/w**2
>>> cp=solve(dC, w);cp
[20**(1/3)*3**(2/3)/3]
>>> FunIncDecS(dC, w, cp)[0]
         0    1
0   1.38    -
1   1.88    0
2   2.38    +
>>> ddc=diff(C, w, 2)
>>> [N(ddc.subs(w, i), 3) for i in [1.78, 1.88, 1.98]]
[4.04e+3, 3.61e+3, 3.26e+3]
```

As a result, C'(w) tends to decrease and increase with respect to the critical point, and C''(w)>0 indicates that C(w) is concave up.

Therefore, it becomes the relative minimum at the critical point, and C(critical point) becomes the minimum value of the price.

```
>>> C.subs(w, cp).evalf(5)
6376.0
```

Ex) Create a cylindrical can that can hold 1.5 L of liquid. Determine the dimensions of the cans to use the least material?

height: h, radius: r

Volume $v=\pi r^2 h=1.5 \rightarrow h=1.5/(\pi r^2)$

The minimum material requirement is equivalent to minimizing the surface area of the can.

Suface Area: $A=2\pi rh+2\pi r^2$

```
>>> h, r=symbols('h, r', real=True)
>>> h=1.5/(2*pi*r**2)
>>> A=2*pi*r*h+2*pi*r**2;A
2*pi*r**2 + 1.5/r
>>> dA=diff(A, r);dA
4*pi*r - 1.5/r**2
>>> cp=solve(dA, r);cp
[0.492372510921348]
>>> FunIncDecS(dA, r, cp)[0]
          0      1
0 -0.00763    -
1  0.492       0
2  0.992       +
>>> ddA=diff(A, r, 2);ddA
```

4*pi + 3.0/r**3

>>> [N(ddA.subs(r, i), 3) for i in [0.490, 0.492, 0.5]]

[38.1, 37.8, 36.6]

Around critical point, A'(r) is swifted to '+' from '-' and A''(r)>0 (concave up). Therefore, the local minimum height and area are determined to at the critical point.

>>> h.subs(r, cp).evalf(5) #h

0.98475

>>> A.subs(r, cp).evalf(5)

4.5697

Ex) The window shown in the following figure should be made of a total of 12 m of material. Determining the dimensions of the window to make it the largest?

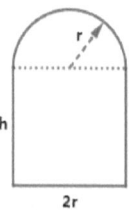

Area of window: A=1/2 πr²+2hr

condition: 12=2h+2r+πr

>>> r, h=symbols("r, h", real=True)
>>> h=solve(2*h+2*r+pi*r-12, h);h
[-pi*r/2 - r + 6]

115

```
>>> A=(1/2)*pi*r**2+2*h[0]*r;A
 0.5*pi*r**2 + r*(-pi*r - 2*r + 12)
>>> dA=diff(A, r);dA
 r*(-pi - 2) - 2*r + 12
>>> cp=solve(dA, r);cp
 [12/(pi + 4)]
>>> FunIncDecS(dA, r, cp)[0]
       0    1
0 1.18  +
1 1.68  0
2 2.18  -
>>> ddA=diff(A, r, 2);ddA
-(pi + 4)
>>> [N(ddA.subs(r, i), 3) for i in [1.58, cp[0], 1.78]]
[-7.14, -7.14, -7.14]
```

Around critical point, the sign of A'(r): '+' →'-', A''(r)>0: concave up

A(critical point): local Maximum

```
>>> A.subs(r, cp[0]).evalf(5)
 10.082
>>> h[0].subs(r, cp[0]).evalf(5)
 1.6803
```

Ex) The largest area of the rectangular in a circle (radius =4)?

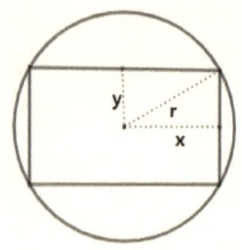

Area: A=4xy

condition: $x^2+y^2=16$, $0 \leq x \leq 4$

```
>>> x, y, h=symbols("x, y, h", real=True)
>>> y=solve(x**2+y**2-16, y);y
[-sqrt(-x**2 + 16), sqrt(-x**2 + 16)]
```

Because y≥0, y=y[1].

```
>>> A=4*x*y[1];A
4*x*sqrt(-x**2 + 16)
>>> A=4*x*y[1];A
4*x*sqrt(-x**2 + 16)
>>> dA=diff(A, x);dA
-4*x**2/sqrt(-x**2 + 16) + 4*sqrt(-x**2 + 16)
>>> cp=solve(dA, x);cp
[-2*sqrt(2), 2*sqrt(2)]
```

Because x≥0,cp=cp[1].

```
>>> FunIncDecS(dA, x, [cp[1]], 0.1)[0]
        0    1
0 2.73   +
```

1 2.83   0

2 2.93   -

3 3.03   -

```
>>> ddA=diff(A, x, 2);ddA
>>> [N(ddA.subs(x, i), 3) for i in [cp[1]-0.1, cp[1], cp[1]+1]]
[-14.4, -16.0, -184.]
```

From the result, A(x) has the local maximum at critical point.

```
>>> A.subs(x, cp[1])
32
```

Ex) Determine the point closest to (0, 2) out of the points on the curve of y=x²+1?

If the point on the curve is (x, y), the distance between the point and (0, 2) can be calculated as follows.

$$d = \sqrt{(x-0)^2+(y-2)^2} = \sqrt{x^2+(y-2)^2}$$

Determine the local minimum value of d.

```
>>> x, y=symbols("x, y", real=True)
>>> y=x**2+1
>>> D=(x**2+(y-2)**2)**0.5;D
(x**2 + (x**2 - 1)**2)**0.5
>>> dD=diff(d, x);dD
(x**2 + (x**2 - 1)**2)**(-0.5)*(2.0*x*(x**2 - 1) + 1.0*x)
>>> cp=solve(dD, x);cp
[-0.707106781186548, 0.0, 0.707106781186548]
>>> FunIncDecS(dD, x, cp)[0]
          0                1
```

0	-1.21	-
1	-0.707	0
2	-0.207	+
3	0	0
4	0.293	-
5	0.707	0
6	0.793	+
7	1.29	+

```
>>> y.subs(x, cp[0])
1.50000000000000
>>> y.subs(x, cp[2])
1.50000000000000
```

From the Result,

(-0.7, 1.5), (0.7, 1.5)

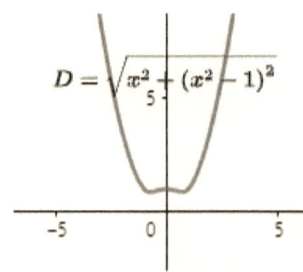

$$D = \sqrt{\frac{x^2}{5} + (x^2 - 1)^2}$$

2 Integral

2.1 Indefinite integral

2.1.1 Introduction

Here is a function generated by differentiating a function. Think about the original function.

$f(x)=x^4+3x-9$

The derivative is calculated using the following formula:

$(x^n)'=nx^{n-1}$

Applying the above formula to the function f(x) produces the following function.

$$\left(\frac{1}{5}x^5\right)'=x^4, \quad \left(\frac{3}{2}x^2\right)'=3x, \quad (-9x)'=-9$$

$$\rightarrow \left(\frac{1}{5}x^5+\frac{3}{2}x^2-9x\right)' = x^4+3x-9$$

If the above function f(x) is differentiated, the constant term becomes 0. This means that integrating f'(x), which is a differential function of f(x), means that the originally existing constant is not reflected in the integrated function. Therefore, if a function f(x) is differentiated to f'(x), then an integral constant 'c' must be added to the integral function f(x).

$$F(x)=\frac{1}{5}x^5+\frac{3}{2}x^2-9x+c, \quad c:\text{constant}$$

If the integral function of the function f(x) is F(x) then the following relation holds:

F'(x)=f(x)

The integral is indicated by the "∫" symbol. Therefore, the integral of f(x) is expressed as

$$\int f(x)dx=F(x)+c \quad c:\text{constant}$$

The integral of the state where the constant can not be determined as above is called an indefinite integral. The general formula of the integral is as follows.

$$\int x^n dx=\frac{1}{n+1}x^{n+1}+c \quad c:\text{constant}, \ n\neq 1$$

Integrals of certain functions such as trigonometric functions and logarithmic functions are computed by applying the derivative formula backwards.

When the above formula is applied to the integral of the function f(x):

$$\int x^4+3x-9dx=\frac{1}{4+1}x^{4+1}+\frac{3}{1+1}x^{1+1}-\frac{9}{1}x+c=\frac{1}{5}x^5+\frac{3}{2}x^2-9x+c$$

Execute the integration using <u>integrate()</u> function of the sympy module. Note that the indefinite integral by the integrate() function does not include a constant term.

```
>>> from sympy import *
>>> x=symbols('x')
>>> f=x**4+3*x-9;f
x**4 + 3*x - 9
>>> integrate(f, x)
x**5/5 + 3*x**2/2 - 9*x
```

An indefinite integral has the following characteristics:

1) $\int kf(x)dx = k\int f(x)dx$ k:constant

However, it is not possible to multiply two functions as follows.

2) $\int f(x)g(x)dx \neq \int f(x)dx \int g(x)dx$

3) $\int kdx = kx + c$ k, c:constant

```
>>> f=x**4;f
x**4
>>> integrate(3*f, x)
3*x**5/5
>>> 3*integrate(f, x)
3*x**5/5
>>> g=2*x**3;g
2*x**3
>>> integrate(f, x)+integrate(g, x)
x**5/5 + x**4/2
```

```
>>> integrate(f+g, x)
x**5/5 + x**4/2
>>> integrate(f, x)+integrate(g, x)==integrate(f+g, x)
True
>>> integrate(f, x)*integrate(g, x)
x**9/10
>>> integrate(f*g, x)
x**8/4
>>> integrate(f, x)*integrate(g, x)==integrate(f*g, x)
False
```

Ex) Integrate the followings?

```
>>> x, y, w=symbols('x, y, w', real=True)
```

1) $\int (5x^3-10x^{-6}+4)\ dx = \dfrac{5}{4}x^4-\dfrac{10}{-5}x^{-5}+4x+c$

$$= \dfrac{5}{4}x^4+2x^{-5}+4x+c$$

```
>>> integrate(5*x**3-10*x**(-6)+4, x)
5*x**4/4 + 4*x + 2/x**5 +c
>>> integrate(1, y)
y
```

3) $\int \dfrac{w^2+\sqrt[3]{w}}{w^2}\ dw = \int \dfrac{w^2}{w^2}\ dw+\int \dfrac{\sqrt[3]{w}}{w^2}\ dw$

$$= w+\int w^{-\frac{5}{3}}\ dw$$

$$= w-\dfrac{3}{2}w^{-\frac{2}{3}}+c$$

```
>>> integrate((w**2+w**Rational("1/3"))/w**2, w)
w - 3/(2*w**(2/3))
```

4) $\int 3e^x+5\cos(x)-10\sec^2(x)\ dx = 3e^x+5\sin(x)-10\tan(x)+c$

$\leftarrow (e^x)' = e^x,\ (\sin(x))' = \cos(x)$

$(\tan(x))' = \left(\dfrac{\sin(x)}{\cos(x)}\right)' = \dfrac{\cos(x)\cos(x)+\sin(x)\sin(x)}{\cos^2(x)} = \dfrac{1}{\cos^2(x)} = \sec^2(x)$

The functions relate to the integration of exponential and trigonometric functions and do not apply the general formula of integral. (See Integrating trigonometric and exponential functions)

```
>>> f=integrate(3*exp(x)+5*cos(x)-10*sec(x)**2, x);f
3*exp(x) + 5*sin(x) - 10*sin(x)/cos(x)
>>> simplify(f)
3*exp(x) + 5*sin(x) - 10*tan(x)
```

Ex) Calculate the function f (x) according to the condition.

1) $f'(x)=4x^3-9+2\sin(x)+7e^x, f(0)=15$

$f(x)=\int 4x^3-9+2\sin(x)+7e^x\ dx = x^4-9x-2\cos(x)+7e^x+c$

$f(0)=-2+7+c=15 \rightarrow c=10$

$\therefore\ f(x)=x^4-9x-2\cos(x)+7e^x+10$

```
>>> f=integrate(4*x**3-9+2*sin(x)+7*exp(x), x);f
x**4 - 9*x + 7*exp(x) - 2*cos(x)
f(0)=7 - 2cos(0)+c=5+c=15
>>> f0=(f+c).subs(x, 0);f0
c + 5
```

124

$c+5=15 \rightarrow c=10$

Use the solve() function to compute the solution of c in the expression c+5-15=0.

```
>>> c1=solve(f0-15, c);c1
[10]
>>> f=f+c1[0];f
x**4 - 9*x + 7*exp(x) - 2*cos(x) + 10
```

2) f(x) from $f"(x)=15x^{1/2}+5x^3+6$, f(1)=-5/4, f(4)=404?

```
>>> x, a, b=symbols('x, a, b')
>>> intf=integrate(15*x**0.5+5*x**3+6, x);intf
 5*x**4/4 + 6*x + 10.0*x**1.5
>>>intf1=intf+a; intf1
a + 5*x**4/4 + 6*x + 10.0*x**1.5
>>>F1=integrate(intf1, x);F1
a*x + x**5/4 + 3*x**2 + 4.0*x**2.5
>>> F=F1+b;F1
a*x + x**5/4 + 3*x**2 + 4.0*x**2.5
```

There are two variables a and b. Create a simultaneous equation from f(1)=-5/4, f(4)=404 and calculate the solution. This process can be applied to the solve() function.

```
>>> eq1=F.subs(x, 1)+(5/4);eq1
a + b + 8.5
>>> eq2=F.subs(x, 4)-404;eq2
4*a + b + 28.0
```

```
>>> sol=solve([eq1, eq2], a, b);sol
{a: -6.50000000000000, b: -2.00000000000000}
>>> F=F.subs({a:round(sol[a],1), b:round(sol[b], 1)});F
x**5/4 + 3*x**2 - 6.5*x + 4.0*x**2.5 - 2.0
```

2.1.2 Substitution Rule

The following complex functions can be simplified by replacing certain parts with other variables.

$$\int 18x^2 \sqrt[4]{6x^3+5}\ dx$$

$6x^3+5=u,$ By $Differential$ of both sides

$\rightarrow 18x^2dx=du$

$$\int \sqrt[4]{u}\ du=\frac{4}{5}u^{\frac{5}{4}}+c=\frac{4}{5}(6x^3+5)^{\frac{5}{4}}+c$$

In the above procedure, $6x^3 + 5$ has been replaced by u. It is possible to simplify the expression because the derivative of the substituted part is contained in the original function. This method is called substitution, and the integration by this technique is called the substitution integration.

```
>>>f=18*x**2*(6*x**3+5)**Rational(1, 4);f
18*x**2*(6*x**3 + 5)**(1/4)
>>> F=integrate(f, x);F
24*x**3*(6*x**3 + 5)**(1/4)/5 + 4*(6*x**3 + 5)**(1/4)
>>> simplify(F)
4*(6*x**3 + 5)**(5/4)/5
```

Using the sympy functions, the substitution integral is performed as follows.

1) Variable (symbol) definition and function creation: symbols(), f

2) Variable substitution: f1=fs ({x: u, y: diff(u)})

3) Integrating the substituted expression: FU=integrate(f1, u)

4) Reduction of the substituted part: FU.subs(u, x)

However, substitution integrals can be added to other processes in addition to the above four steps.

```
>>>x, u=symbols("x, u")
>>> f=18*x**2*(6*x**3+5)**Rational(1, 4);f
18*x**2*(6*x**3 + 5)**(1/4)
>>> f1=f.subs({(6*x**3 + 5):u, 18*x**2:diff(u, u)});f1
u**(1/4)
>>> F1=integrate(f1, u);F1
4*u**(5/4)/5
```

The above replacement process is summarized as follows.

$$\int f(g(x))g'(x)dx = \int f(u)du$$
$$\because u=g(x), \quad du=g'(x)dx$$

Ex) Integrate the following ?

1) $\int \left(1-\dfrac{1}{w}\right)\cos(w-\ln(w))\,dw$

$v=w-\ln(w) \;\longrightarrow\; dv=\left(1-\dfrac{1}{w}\right)dw$

$\int \cos(v)\,dv = \sin(v)+c = \sin(w-\ln(w))+c$

```
>>> w, v=symbols("w, v")
>>> f=(1-1/w)*cos(w-ln(w));f
(1 - 1/w)*cos(w - log(w))
>>> f1=f.subs({w-ln(w):v, (1-1/w):diff(v, v)});f1
cos(v)
>>> F1=integrate(f1, v);F1
sin(v)
>>> F=F1.subs(v, w-ln(w)); F
sin(w - log(w))
```

It is the same as the result of integrating without applying the substitution integral.

```
>>> F=integrate(f, w);F
sin(w - log(w))
```

2) $\int \dfrac{x}{\sqrt{1-4x^2}}\,dx$

$u=1-4x^2 \;\longrightarrow\; du=-8x\,dx, \quad \dfrac{1}{-8}du=dx$

$\dfrac{-1}{8}\int \dfrac{1}{\sqrt{u}}\,du=-\dfrac{2}{8}\sqrt{u}=-\dfrac{1}{4}\sqrt{1-4x^2}+c$

```
>>> x, u=symbols("x, u")
>>> f=x/(sqrt(1-4*x**2));f
```

x/sqrt(-4*x**2 + 1)

>>> f1=f.subs({1-4*x**2:u, x:-diff(u, u)/8});f1

-1/(8*sqrt(u))

>>> F1=integrate(f1, u);F1

-sqrt(u)/4

>>> F=F1.subs(u, 1-4*x**2); F

-sqrt(-4*x**2 + 1)/4

>>> Fdir=integrate(f, x);Fdir

-sqrt(-4*x**2 + 1)/4

3) $\int \sin(1-x)(2-\cos(1-x))^4\,dx$

$u=2-\cos(1-x),\ du=-\sin(1-x)\,dx$

$-\int u^4\,du = -\dfrac{1}{5}u^5+c =- \dfrac{1}{5}(2-\cos(1-x))^5+c$

>>> g=sin(1-x)*(2-cos(1-x))**4;g

-(-cos(x - 1) + 2)**4*sin(x - 1)

>>>g1=g.subs({2-cos(1-x):u, sin(1-x):-diff(u, u)}});g1

-u**4

>>> G1=integrate(g1);G1

-u**5/5

>>> G=G1.subs(u, 2-cos(1-x));G

-(-cos(x - 1) + 2)**5/5

Integrating function g directly using integrate() returns a complex result.

>>> Gdir=integrate(g, x);simplify(Gdir)

2*sin(x - 1)**4 + 4*sin(x - 1)**2*cos(x - 1)**2 + cos(x - 1)**5/5 +

8*cos(x - 1)**3 - 16*cos(x - 1)**2 + 16*cos(x - 1)

Integral means the area occupied by the function (See definite section). In this sense, each result function (G & Gdir) returns the same value when calculating the difference between the values of a certain interval.

```
>>>N(G.subs(x, 1)-G.subs(x, 0), 3)
1.13
>>>N(Gdir.subs(x, 1)-Gdir.subs(x, 0), 3)
1.13
```

4) $\int \dfrac{3x}{(5x^2+4)^2}\, dx$

$u = 5x^2+4,\ du = 10x\, dx \rightarrow \dfrac{1}{10}du = x\, dx$

$\dfrac{3}{10}\int \dfrac{1}{u^2}du = -\dfrac{3}{10}\dfrac{1}{u}+c = -\dfrac{3}{10}\dfrac{1}{5x^2+4}+c = -\dfrac{3}{50x^2+40}+c$

```
>>> x,u=symbols("x,u")
>>>g=3*x/(5*x**2+4)**2;g
3*x/(5*x**2 + 4)**2
>>>gu=g.subs({5*x**2+4:u, x:diff(u, u)/10});g1
3/(10*u**2)
>>>Gu=integrate(gu, u);Gu
-3/(10*u)
>>> G=Gu.subs(u,5*x**2+4);G
-3/(10*(5*x**2 + 4))
>>> simplify(G)
```

-3/(50*x**2 + 40)

>>> Gdir=integrate(g, x);Gdir

-3/(50*x**2 + 40)

5) $\int \dfrac{2x^3+1}{x^4+2x}\,dx$

$u=x^4+2x,\ du=(4x^3+2)dx \rightarrow \dfrac{1}{2}du=(2x^3+1)dx$

$\dfrac{1}{2}\int \dfrac{1}{u}\,du = \dfrac{1}{2}\ln(|u|)+c = \dfrac{1}{2}\ln(|x^4+2|)+c$

>>> x,u=symbols("x,u")

>>>g=(2*x**3+1)/(x**4+2*x);g

(2*x**3 + 1)/(x**4 + 2*x)

>>> t=x**4+2*x;t

x**4 + 2*x

>>> t1=simplify((2*x**3+1)/diff(t,x));t1

1/2

>>> gu=g.subs({t:u, 2*x**3+1:t1});gu

1/(2*u)

>>> GU=integrate(gu, u);GU

log(u)/2

>>> G=GU.subs(u, t);G

log(x**4 + 2*x)/2

>>> Gdir=integrate(g, x);Gdir

log(x**4 + 2*x)/2

6) $\int \left(x\cos(x^2+1) + \dfrac{x}{x^2+1} \right) dx = \int x\cos(x^2+1)dx + \int \dfrac{x}{x^2+1}\, dx$

$u=x^2+1,\ du=2x\,dx \rightarrow \dfrac{1}{2}du=x\,dx$

$\dfrac{1}{2}\int \cos(u)du + \dfrac{1}{2}\int \dfrac{1}{u}du = \dfrac{1}{2}\sin(u) + \dfrac{1}{2}\ln(u)+c$

$$= \dfrac{1}{2}\sin(x^2+1) + \dfrac{1}{2}\ln(x^2+1)+c$$

```
>>>x,u=symbols("x,u")
>>>g=x*cos(x**2+1)+x/(x**2+1);g
x*cos(x**2 + 1) + x/(x**2 + 1)
>>>t=x**2+1;t
x**2 + 1
>>>t1=simplify(x/diff(t,x));t1
1/2
>>>gu=g.subs({t:u, x:t1});gu
cos(u)/2 + 1/(2*u)
>>>GU=integrate(gu, u);GU
log(u)/2 + sin(u)/2
>>>G=GU.subs(u, t);G
log(x**2 + 1)/2 + sin(x**2 + 1)/2

>>>Gdir=integrate(g, x);Gdir
log(x**2 + 1)/2 + sin(x**2 + 1)/2
```

2.2 Definite Integral

2.2.1 Introduction

Differential and integral of a function are contrasted with each other. Derivative is to decompose a function into a minute part to see the characteristics of the function, but the integration is applied to derive the overall characteristic by adding a small section of the function. Integrating means to add a small part of a function so that the area occupied by the function in a certain interval can be calculated.

For example, as shown in Figure 2.1, the area of [0, 2] of f (x) = x^2 + 1 is calculated as follows.

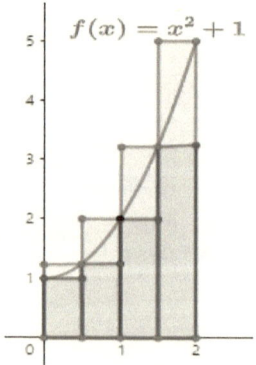

Figure 2.1 Area by the function in [0, 2].

133

Figure 2.1 divides the [0,2] section into four small squares. Each rectangle has the same x-axis length and its y-axis length is a function f(x), so the area of each rectangle can be easily calculated. The sum of each rectangle will approximate the total area occupied by the function.

The x-length of the top rectangle is (2-0) / 4 = 1/2.

As shown in Figure 2.1, rectangles can be divided into two types (Red and blue squares). First, calculate the area based on the red rectangle.

$$A = \frac{1}{2}f(0.5) + \frac{1}{2}f(1) + \frac{1}{2}f(1.5) + \frac{1}{2}f(2)$$

```
>>> x=symbols("x")
>>> f=x**2+1;f
x**2 + 1
>>> bottom=(2-0)/4;bottom
0.5
>>> A=sum([bottom*f.subs(x, i) for i in [0.5, 1, 1.5, 2]]);A
5.75000000000000
```

The sum of areas separated by a blue line is also calculated.

$$A = \frac{1}{2}f(0) + \frac{1}{2}f(0.5) + \frac{1}{2}f(1) + \frac{1}{2}f(1.5)$$

```
>>> A=sum([bottom*f.subs(x, i) for i in [0, 0.5, 1, 1.5]]);A
3.75000000000000
```

There is a big difference between the results of applying the blue squares and the red squares. However, as the number of squares contained in the curve increases, the difference between the two parts will be smaller and closer to the area due to the function. If it is divided into 100 sections in the same way as above, it is as follows.

```
>>> bottom=(2-0)/100;bottom
0.02
>>> rng=np.append(np.arange(0, 2, bottom),2);rng
array([0. , 0.02, ..., 2. ])
>>> A1=sum([bottom*f.subs(x, rng[i]) for i in range(1, len(rng))]);A1
4.70680000000000
>>> A2=sum([bottom*f.subs(x, rng[i]) for i in range(len(rng)-0)]);A2
4.72680000000000
```

That is, increasing the number of intervals results in the same result for both methods.

$$A \approx \sum_{i=1}^{n} f(x_i)\Delta x \rightarrow A = \lim_{n \to \infty} \sum_{i=1}^{n} f(x_i)\Delta x$$

<Definite Integral>

If the function f(x) is continuous in the interval [a, b], the interval can be equally divided by a very small interval Δx.

The area of the total section can be calculated by calculating the height f(x*i) of a specific part of the section.

This form is called the definite integral from section a(lower limit) to section b(upper limit), as shown below.

$$\int_a^b f(x)\,dx = \lim_{n\to\infty} \sum_{i=1}^{n} f(x_i^*) = F(x)\big|_a^b = F(b) - F(a)$$

a: lower limit

b: upper limit

To calculate the area of the area occupied by the function as shown in Figure 2.1, the definite integral is applied as follows.

$$\int_0^2 (x^2+1)\,dx = \frac{1}{3}x^3 + 1\bigg|_0^2 = \frac{1}{3}(8+2) = \frac{14}{3}$$

```
>>> integrate(x**2+1, (x, 0, 2))
14/3
>>> N(integrate(x**2+1, (x, 0, 2)),3)
4.67
```

<Characteristic of integral>

1) $\displaystyle\int_a^b f(x)\,dx = -\int_b^a f(x)\,dx$

2) $\displaystyle\int_a^a f(x)\,dx = 0 \Rightarrow \lim_{n\to\infty} f(x_i^*)\frac{a-a}{n} = 0$

3) $\displaystyle\int_a^b cf(x)\,dx = c\int_a^b f(x)\,dx$

4) $\displaystyle\int_a^b f(x)\pm g(g)\,dx = \int_a^b f(x)\,dx \pm \int_a^b g(x)\,dx$

$\Rightarrow \displaystyle\lim_{n\to\infty} \sum_{i=1}^{n}(f(x_i^*)\pm g(x_i^*)) = \lim_{n\to\infty} \sum_{i=1}^{n} f(x_i^*)\frac{b-a}{n} \pm \lim_{n\to\infty} \sum_{i=1}^{n} g(x_i^*)\frac{b-a}{n}$

5) $\displaystyle\int_a^b f(x)\,dx = \int_a^c f(x)\,dx \pm \int_c^b f(x)\,dx, \quad$ c:real number

c does not necessarily have to be in the interval [a, b].

$$6) \int_a^b f(x)dx = \int_a^b f(t)dt$$

$$7) \int_a^b cdx = c(b-a), \quad c: \text{constant}$$

$$\Rightarrow \lim_{n\to\infty} \sum_{i=1}^{n} c\frac{b-a}{n} = \lim_{n\to\infty} cn\frac{b-a}{n} = \lim_{n\to\infty} c(b-a) = c(b-a)$$

$$8) \ a \leq x \leq b, \ f(x) \geq 0 \Rightarrow \int_a^b f(x)dx \geq 0$$

$$9) \ a \leq x \leq b, \ f(x) \geq g(x) \Rightarrow \int_a^b f(x)dx \geq \int_a^b g(x)dx$$

$$10) \ a \leq x \leq b, \ m \leq f(x) \leq M \Rightarrow m(b-a) \leq \int_a^b f(x)dx \leq M(b-a)$$

$$11) \ \left| \int_a^b f(x)dx \right| \leq \int_a^b |f(x)|dx$$

Formula 11 above is described as follows.

$$|a| \leq b \rightarrow -b \leq a \leq b$$

Therefore, the following equation holds.

$$|f(x)| \leq f(x) \Rightarrow -|f(x)| \leq f(x) \leq |f(x)|$$

$$\left| \int_a^b f(x)dx \right| \leq \int_a^b |f(x)|dx \Rightarrow -\int_a^b |f(x)|dx \leq \left| \int_a^b f(x)dx \right| \leq \int_a^b |f(x)|dx$$

Ex) Integrate the followings?

$$1) \int_2^0 x^2+1\,dx = \int_0^2 x^2+1\,dx$$

$$\rightarrow \int_2^0 x^2+1\,dx = \frac{1}{3}x^3+x\Big|_2^0 = -\frac{14}{3}$$

$$\rightarrow -\int_0^2 x^2+1\,dx = -\left(\frac{1}{3}x^3+x\right)\Big|_2^0 = -\frac{14}{3}$$

>>> f=x**2+1

>>> integrate(f, (x, 2, 0))

-14/3

```
>>> integrate(-f, (x,0, 2))
-14/3
```

2) $\int_0^2 10x^2+10\,dx$

$$= 10\int_0^2 x^2+1\,dx = 10\left(\frac{1}{3}x^3+x\right)\Big|_0^2 = \frac{140}{3}$$

```
>>> f=10*x**2+10;f
10*x**2 + 10
>>> integrate(f, (x, 0, 2))
140/3
```

$g(x)=\int_{-4}^x t^3+t^2+1\,dt$

$$= \frac{t^4}{4}+\frac{t^3}{3}+t\Big|_{-4}^x$$

$$=\frac{x^4}{4}+\frac{x^3}{3}+x-\left(\frac{(-4)^4}{4}+\frac{(-4)^3}{3}+(-4)\right)$$

$g'(x)=x^3+x^2+1$

$g'(x)=\left(\int_{-4}^x t^3+t^2+1\,dt\right)=x^3+x^2+1$

$\therefore g(x)=\int g'(x)\,dt$

```
>>> x, t=symbols("x, t")
>>> f=x**2+1;f
x**2 + 1
>>> integrate(f, (x, 0, 2))
14/3
>>> g=t**2+1;g
t**2 + 1
>>> Integrate(g, (t, 0,2))
14/3
```

Ex) Integrate the [-5, 12] section of the function with the following result.

$$\int_{12}^{-10} f(x)dx=6, \int_{100}^{-10} f(x)dx=-2, \int_{100}^{-5} f(x)dx=4$$

$$\int_{-5}^{12} f(x)dx?$$

$$= \int_{-5}^{100} f(x)dx + \int_{100}^{-10} f(x)dx + \int_{-10}^{12} f(x)dx$$

$$= -4-2-6$$

$$= -12$$

Ex) Integrate the followings ?

$$f(x)=\begin{cases} 6 & \text{if } x\leq 1 \\ 3x^2 & \text{if } x>1 \end{cases}$$

$$\int_{-2}^{3} f(x)dx = \int_{1}^{3} 3x^2dx + \int_{-2}^{1} 6dx = x^3|_{1}^{3} +6x|_{-2}^{1} =26+18=44$$

>>> integrate(6, (x, -2, 1))+integrate(3*x**2,(x, 1, 3))
44

Ex) Integrate the followings ?

$$\int_{0}^{3} |3t-5|dt$$

The absolute function is divided into negative and positive divisions.

$$3t-5\geq0 \rightarrow t \geq 5/3 \Rightarrow f(t)=3t-5$$
$$3t-5<0 \rightarrow t < 5/3 \Rightarrow f(t)=5-3t$$

$$\int_0^3 |3t-5|\,dt$$

$$= \int_{\frac{5}{3}}^3 (3t-5)\,dt + \int_0^{\frac{5}{3}} (5-3t)\,dt$$

$$= \left. \frac{3}{2}t^2 - 5t \right|_{\frac{5}{3}}^{3} + \left. 5t - \frac{3}{2}t^2 \right|_0^{\frac{5}{3}}$$

$$= \frac{41}{6}$$

```
>>> integrate(3*t-5, (t, Rational("5/3"), 3))+integrate(5-3*t, (t, 0,
Rational("5/3")))
41/6
```

Rational(x): Keeps the form of fraction x

2.2.2 Interpretation of definite integral

The result of a definite integral can be interpreted as the area occupied by a function of a particular interval, as mentioned above. Such a meaning is a net change by a function in a certain interval. For example, if the volume function of water per hour filled in the water tank is V (t), then the integration of [a, b] time is as follows.

$$\int_a^b V'(t)\,dt = V(b) - V(a)$$

The above calculation result means the following.

1) Volume of water filled in the water tank at time b-a

2) net change in water between specific times in tank (a-b)

Ex) Differentiate the followings?

$$g(x)= \int_{-4}^{x} t^3+t^2+1\, dt$$

$$= \frac{t^4}{4} + \frac{t^3}{3} + t \Big|_{-4}^{x}$$

$$= \frac{x^4}{4} + \frac{x^3}{3} + x - \left(\frac{(-4)^4}{4} + \frac{(-4)^3}{3} + (-4) \right)$$

$$g'(x) = x^3 + x^2 + 1$$

$$g'(x) = \left(\int_{-4}^{x} t^3 + t^2 + 1\, dt \right)' = x^3 + x^2 + 1$$

$$\therefore g(x) = \int g'(x)\, dt$$

In integrals, upper and low bound are variables and constants, respectively. Therefore, the result is a combination of variables and constants. In the derivative of this result, the constant term is 0, so only the variable term exists. As a result, the derivative of the integrated result is equivalent to assigning a new variable to the function to integrate.

```
>>> x,t=symbols('x,t')
>>> f=t**3+t**2+1;f
t**3 + t**2 + 1
>>>F=integrate(f, (t, -4, x));F
x**4/4 + x**3/3 + x - 116/3
>>> diff(F, x)
x**3 + x**2 + 1
```

In the above example, the following relationship is established:

If the function f(x) is continuous in the interval [a, b] and the integral function up to an arbitrary point x in the interval is g(x), it can be expressed as follows.

$$g(x)= \int_a^x f(t)\, dt \rightarrow g'(x)=f(x)\, dx$$

<Proof>

If a <x <x + h <b, the following holds:

$$g(x+h)-g(x)= \int_a^{x+h} f(x)dx - \int_x^a f(x)dx = \int_x^{x+h} f(x)dx$$

If h≠0,

$$\frac{g(x+h)-g(x)}{h} = \frac{1}{h}\int_x^{x+h} f(x)dx \quad (1)$$

The left side of equation(1) represents the rate of change between x and x+h and can be said to be between the minimum and maximum values in that interval. The right term represents the area by function of the interval. Therefore, in the condition of x≤c≤d≤x+h, equation(1) can be expressed as follows. (Figure 2.2)

$$h\cdot f(c) \leq \int_x^{x+h} f(x)dx \leq h\cdot f(d)$$

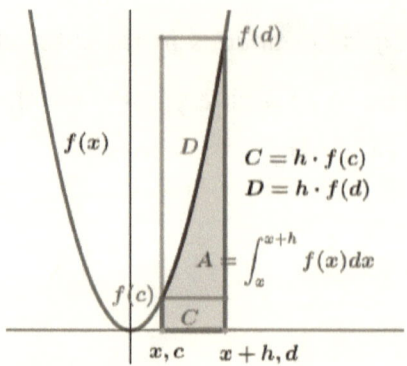

Figure 2.2. Integral of function f(x) in [c, d].

$$f(c)\leq\frac{1}{h}\int_x^{x+h} f(x)dx\leq f(d)$$

$$\rightarrow f(c)\leq\frac{g(x+h)-g(x)}{h}\leq f(d) \quad (2)$$

From the condition, h→0⇒c→x, d→x. Because x<c<d<x+h.

$$\lim_{h\to 0}\frac{g(x+h)-g(x)}{h}=g'(x)=f(x)$$

$$\lim_{h\to 0} f(c)=\lim_{c\to x} f(c)=f(x)$$

$$\lim_{h\to 0} f(d)=\lim_{d\to x} f(d)=f(x)$$

Therefore, Equation (2) is as follows.

$$f(x)\leq\frac{g(x+h)-g(x)}{h}=g'(x)\leq f(x)\Rightarrow g'(x)=f(x)$$

Ex) Differentiate the followings?

143

$$\int_{x^2}^{1} \frac{t^4+1}{t^2+1} dt = -\int_{1}^{x^2} \frac{t^4+1}{t^2+1} dt$$

$$\frac{d}{dx}\left(-\int_{1}^{x^2} \frac{t^4+1}{t^2+1} dt\right) = -\frac{(x^2)^4+1}{(x^2)^2+1}(x^2)' = -2x\frac{(x^2)^4+1}{(x^2)^2+1}$$

>>> g=integrate((t**4+1)/(t**2+1), (t, x**2, 1));g

-x**6/3 + x**2 - 2*atan(x**2) - 2/3 + pi/2

>>> diff(g, x)

-2*x**5 + 2*x - 4*x/(x**4 + 1)

>>> factor(diff(g, x))

-2*x*(x**8 + 1)/(x**4 + 1)

For the above problem, the chain rule (y'= f'(g(x))g'(x)) has been applied.

If the upper bound of the integral is a variable, the derivative of the integral function is calculated as:

$$\frac{d}{dx}\left(\int_{a}^{u(x)} f(t)dt\right) = u'(x)f(u(x))$$

If both the upper and lower limits of the integral are variables, the derivative of the integral function is calculated as follows.

$$\frac{d}{dx}\left(\int_{v(x)}^{u(x)} f(t)\,dt\right)$$

$$= \frac{d}{dx}\left(-\int_{a}^{v(x)} f(t)\,dt + \int_{a}^{u(x)} f(t)\,dt\right)$$

$$= -f(v(x))v'(x) + f(u(x))u'(x)$$

Ex) Differentiate the follows?

$$\int_{\sqrt{x}}^{3x} t\sin(1+t^2)dt = -\int_a^{\sqrt{x}} t\sin(1+t^2)dt + \int_a^{3x} t\sin(1+t^2)dt$$
$$= -(\sqrt{x})'(\sqrt{x}\ \sin(1+x^2)) + (3x)'(3x\sin(1+9x^2))$$
$$= -\frac{1}{2}\sin(1+x^2) + 9x\sin(1+9x^2)$$

```
>>> x,t=symbols("x,t")
>>> f=t*sin(1+t**2);f
t*sin(t**2 + 1)
>>> F=integrate(f, (t,-sqrt(x),3*x));F
cos(x + 1)/2 - cos(9*x**2 + 1)/2
>>> diff(F, x)
9*x*sin(9*x**2 + 1) - sin(x + 1)/2
```

2.2.3 Even and Odd Function

The even function is a function that satisfies the following relation:

  f(-x)=f(x)

Ex) f(x)=x², f(x)=cos(x)

Odd function is a function that satisfies the following relation.

  f(-x)=-f(x)

Ex) f(x)=x³, f(x)=sin(x)

When the absolute values of the upper and lower limits of the integral are the same, considering the characteristics of the even function and the odd function, the integral of each function can be easily calculated as follows.

For the even function

$$\int_{-a}^{a} f(x)dx = \int_{0}^{a} f(x)dx + \int_{-a}^{0} f(x)$$
$$= \int_{0}^{a} f(x)dx + \int_{0}^{a} f(-x)dx$$
$$= \int_{0}^{a} f(x)dx + \int_{0}^{a} f(x)dx$$
$$= 2\int_{0}^{a} f(x)dx$$

For the odd function

$$\int_{-a}^{a} f(x)dx = \int_{0}^{a} f(x)dx + \int_{-a}^{0} f(x)$$
$$= \int_{0}^{a} f(x)dx + \int_{0}^{a} f(-x)dx$$
$$= \int_{0}^{a} f(x)dx - \int_{0}^{a} f(x)dx$$
$$= 0$$

Ex) Integrate the following?

1) $\int_{-2}^{2} 4x^4 - x^2 + 1\, dx = \int_{-2}^{2} f(x)\, dx$
   $f(-1) = 4 - 1 + 1 = 4$
   $f(1) = 4 - 1 + 1 = 4$
   $\rightarrow f(x) : \text{Even function}$
   $2\int_{0}^{2} 4x^4 - x^2 + 1\, dx = 2\left(\frac{4}{5}x^5 - \frac{1}{3}x^3 + x\right)\Big|_{0}^{2} = \frac{748}{15}$

   >>> f=4*x**4-x**2+1
   >>> integrate(f, (x, -2, 2))
   748/15
   >>> 2*(integrate(f, (x, 0, 2)))

146

748/15

2) $\int_{-10}^{10} x^5 + \sin(x)\,dx = \int_{-10}^{10} f(x)\,dx$

$f(-1) = -1 + \sin(-1)$

$f(1) = 1 + \sin(1)$

$\rightarrow f(x)$: Odd function

$\int_{-10}^{10} x^5 + \sin(x)\,dx = \frac{1}{6}x^6 - \cos(x)\Big|_{-10}^{10} = 0$

```
>>> f=x**5+sin(x)
>>> integrate(f, (x, -10, 10))
0
```

Ex) Integrate the following?

The following functions can be made simple by applying the substitution integration method.

1) $\int_{-2}^{0} 2t^2 \sqrt{1-4t^3}\,dt$

$1-4t^3=u, \ -12t^2 dt=du \rightarrow t^2 dt=-\dfrac{1}{12}\,dt$

$t=0 \rightarrow u=1$

$t=-2 \rightarrow u=33$

$-\dfrac{1}{6}\int_{33}^{1} \sqrt{u}\,du = \dfrac{1}{6}\dfrac{2}{3}\sqrt{u^3}\Big|_{1}^{33} = 20.95$

```
>>> t,u=symbols("t,u")
>>> f=2*t**2*sqrt(1-4*t**3);f
2*t**2*sqrt(-4*t**3 + 1)
>>> y=1-4*t**3
>>> dy=diff(y,t);dy
-12*t**2
```

```
>>> ylim=[y.subs(t, -2), y.subs(t, 0)];ylim
[33, 1]
>>> fu=f.subs(y,u)/dy;fu
-sqrt(u)/6
>>> Fu=integrate(fu, (u, ylim[0], ylim[1]));Fu
-1/9 + 11*sqrt(33)/3
>>> N(Fu, 4)
20.95
```

2) $\displaystyle\int_0^{\frac{1}{2}} e^y+2\cos(\pi y)\,dy = \int_0^{\frac{1}{2}} e^y\,dy+\int_0^{\frac{1}{2}} 2\cos(\pi y)\,dy$

$\pi y=u,\ \pi dy=du \longrightarrow dy=\dfrac{1}{\pi}du$

$y=\dfrac{1}{2} \longrightarrow u=\dfrac{1}{2}\pi$

$y=0 \longrightarrow u=0$

$\displaystyle\int_0^{\frac{1}{2}} e^y\,dy+\int_0^{\frac{1}{2}\pi} \dfrac{2}{\pi}\cos(u)\,du= e^y\Big|_0^{\frac{1}{2}} + \dfrac{2}{\pi}\sin(u)\Big|_0^{\frac{1}{2}\pi}$

$$=e^{\frac{1}{2}}-1+\dfrac{2}{\pi}(1-0)$$

$$=1.285$$

```
>>> y=symbols("y")
>>> f=exp(y)+2*cos(pi*y)
>>> integrate(f, (y, 0, 1/2)).evalf(4)
1.285
```

2.3 Integral Technique

2.3.1 Partial integral

For special functions such as exponential functions, trigonometric functions, etc., it is difficult to apply a general integral formula. These functions can be integrated using the derivative of the function in reverse. Also, when these functions are combined with ordinary functions, the integration of those functions is complicated. In this case, the partial method is used.

The following is the integral of a simple exponential function.

$$\int e^x dx = e^x + c$$

```
>>> integrate(exp(x), x)
exp(x)
```

$x \cdot exp(x^2)$, which is a function combined with a general function, can be calculated by applying a substitution integral.

$$\int xe^{x^2} dx = \frac{1}{2} \int e^u du = \frac{1}{2} e^u + c = \frac{1}{2} e^{x^2} + c$$

$$u = x^2, \; du = 2xdx \; \rightarrow \; \frac{1}{2} du = xdx$$

```
>>> integrate(x*exp(x**2), x)
exp(x**2)/2
```

However, in the above substitution integral, the derivative term of the substitution part must be present in the expression. If it does not, the partial method is applied.

The partial integral can be summarized as follows in relation to the product law of the derivative.

<Partial Integral>

$$\frac{d}{dx}(f(x)g(x)) = \left(\frac{d}{dx}f(x)\right)\cdot g(x) + f(x)\cdot\left(\frac{d}{dx}g(x)\right)$$

$$\int (f(x)g(x))'dx = \int f'(x)g(x)dx + \int f(x)g'(x)dx$$
$$\rightarrow f(x)g(x) = \int f'(x)g(x)dx + \int f(x)g'(x)dx$$
$$\rightarrow \int f(x)g'(x)dx = f(x)g(x) - \int f'(x)g(x)dx$$

$$f(x) = u, \ g(x) = v \rightarrow f'(x)dx = du, \ g'(x)dx = dv$$
$$\therefore \int u\,dv = uv - \int v\,du$$

Since the integrate() function has built-in various integration algorithms, it is not necessary for the user to specify the integrating method for integrating functions. However, in some functions, integrating a function directly may require a long computation time or fail to return an answer. In this case, the user must use a substitution or partial division method to transfer a simple function to integrate(). Here is a user-defined function created to use partial minutes.

```
def partIntegralS(u, dv, Symbolvar, loup=0):
        du=diff(u, Symbolvar)
        v=integrate(dv, Symbolvar)
        uv=u*v
        vdu=integrate(v*du, Symbolvar)
        if loup==0:
            re=uv-vdu
            return(re)
        else:
            re=uv-vdu
            re1=re.subs(Symbolvar, loup[1])-re.subs(Symbolvar, loup[0])
            return(re1)
```

Argument;

u, dv: part of original function

Symbolvar: variable using integration

loup: [low bound, upper bound]. 0 in case of indefinite integral

Ex) Integral the following ?

1) $\int xe^{6x}dx$

$= x\dfrac{1}{6}e^{6x} - \int \dfrac{1}{6}e^{6x}dx$

$= \dfrac{x}{6}e^{6x} - \dfrac{1}{36}e^{6x}+c$

$\because u=x, dv=e^{6x} \rightarrow du=dx, v=\dfrac{1}{6}e^{6x}$

>>> x=symbols('x')

```
>>> f=x*exp(6*x);f
x*exp(6*x)
>>> integrate(f, x)
(6*x - 1)*exp(6*x)/36
```

Using partial integral

```
>>> u=x;u
x
>>> dv=exp(6*x);dv
exp(6*x)
>>> u*integrate(dv, x)-integrate(diff(u, x)*integrate(dv, x))
x*exp(6*x)/6 - exp(6*x)/36
```

Using user-define-function "partIntegrals()"

```
>>> partIntegralS(x, exp(6*x), x)
x*exp(6*x)/6 - exp(6*x)/36
```

2) $\int_{1}^{2} xe^{6x}dx = \dfrac{x}{6}e^{6x}\Big|_{1}^{2} - \dfrac{1}{36}e^{6x}\Big|_{1}^{2}$

```
>>> integrate(f, (x,-1,2))
-5*exp(6)/36 + 11*exp(12)/36
```

- part integral

```
>>> uv=u*integrate(dv, x)
>>> uv1=uv.subs(x, 2)-uv.subs(x, 1);uv1
-exp(6)/6 + exp(12)/3
>>> duv1=integrate(diff(u, x)*integrate(dv, x), (x, 1, 2));duv1
-exp(6)/36 + exp(12)/36
```

```
>>> uv1-duv1
-5*exp(6)/36 + 11*exp(12)/36
```

- partIntegralS()

```
>>> partIntegralS(x, exp(6*x), x, [-1,2])
-5*exp(6)/36 + 11*exp(12)/36
```

3) $\int (3t+5)(\cos(\frac{t}{4}))dt?$

$u=3t+5,\ dv=\cos(\frac{t}{4}) \rightarrow du=3dt,\ v=4\sin(\frac{t}{4})$

$\int (3t+5)(\cos(\frac{t}{4}))dt$

$= (3t+5)4\sin(\frac{t}{4})-\int 4\sin(\frac{t}{4})3dt$

$= 4(3t+5)\sin(\frac{t}{4})+48\cos(\frac{t}{5})+c$

```
>>> t=symbols('t')
>>> integrate((3*t+5)*cos(t/4), t)
12*t*sin(t/4) + 20*sin(t/4) + 48*cos(t/4)
```

- part integral

```
>>> u=3*t+5
>>> dv=cos(t/4)
>>> u*integrate(dv, t)-integrate(diff(u,t)*integrate(dv, t), t)
4*(3*t + 5)*sin(t/4) + 48*cos(t/4)
```

- partIntegralS

```
>>> partIntegralS(3*t+5, cos(t/4), t)
4*(3*w + 5)*sin(w/4) + 48*cos(w/4)
```

4) $\int w^2\sin(10w)dw$

   $u=w^2,\ dv=\sin(10w)$

   $\qquad du=2w\,dw$

   $\rightarrow$ $\begin{cases} \\ v=\int\sin(10w)\,dw \end{cases}$: $\quad$ $\dfrac{x=10w,\quad dx=10dw}{\dfrac{1}{10}\int\sin(x)dx=-\dfrac{1}{10}\cos(x)=-\dfrac{1}{10}\cos(10w)}$

   $\int w^2\sin(10w)dw = -w^2\dfrac{1}{10}\cos(w)-\int 2w\left(-\dfrac{1}{10}\cos(10w)\right)dw$

Again, the integral of the second term of the above equation is calculated by the partial fraction.

$\int w\cos(10w)\,dw$

$=\dfrac{1}{10}w\sin(10w)+\dfrac{1}{100}\cos(10w)$

$\rightarrow$ $\int w^2\sin(10w)dw$

$= -w^2\dfrac{1}{10}\cos(w)-\int 2w\left(-\dfrac{1}{10}\cos(10w)\right)dw$

$= -w^2\dfrac{1}{10}\cos(w)+\dfrac{1}{5}\left(\dfrac{1}{10}w\sin(10w)+\dfrac{1}{100}\cos(10w)\right)$

$= -\dfrac{1}{10}w^2\cos(10w)+\dfrac{1}{50}w\sin(10w)+\dfrac{1}{100}\cos(w)+c$

```
>>> w=symbols('w')
>>> integrate(w**2*sin(10*w), x)
w**2*x*sin(10*w)
>>> w=symbols('w')
>>> integrate(w**2*sin(10*w), w)
-w**2*cos(10*w)/10 + w*sin(10*w)/50 + cos(10*w)/500


>>> u=w**2
```

154

```
>>> dv=sin(10*w)
>>> u*integrate(dv, w)-integrate(diff(u,w)*integrate(dv, w), w)
-w**2*cos(10*w)/10 + w*sin(10*w)/50 + cos(10*w)/500


>>> partIntegralS(w**2, sin(10*w), w)
-w**2*cos(10*w)/10 + w*sin(10*w)/50 + cos(10*w)/500
```

5) $\int x\sqrt{x+1}dx$

$$u = x, dv = \sqrt{x+1} \rightarrow du = dx, v = \frac{2}{3}(x+1)^{\frac{3}{2}}$$

$$x\frac{2}{3}(x+1)^{\frac{3}{2}} - \int \frac{2}{3}(x+1)^{\frac{3}{2}}dx = \frac{2}{3}x(x+1)^{\frac{3}{2}} - \frac{4}{15}(x+1)^{\frac{5}{2}} + c$$

```
>>> simplify(integrate(x*sqrt(x+1), x))
2*sqrt(x + 1)*(3*x**2 + x - 2)/15


>>> u=x
>>> dv=sqrt(x+1)
>>> ans=u*integrate(dv, x)-integrate(diff(u,x)*integrate(dv, x), x);ans
2*x*(x + 1)**(3/2)/3 - 4*(x + 1)**(5/2)/15
>>> simplify(ans)
2*(x + 1)**(3/2)*(3*x - 2)/15


>>> ans1=partIntegralS(x, (x+1)**(1/2), x)
>>> nsimplify(ans1, rational=True)
2*x*(x + 1)**(3/2)/3 - 4*(x + 1)**(5/2)/15
```

6) $\int x^4 e^{\frac{x}{2}} dx$

$=x^4 2e^{\frac{x}{2}} - \int 2e^{\frac{x}{2}} 4x^3 dx$

$u=x^4, \ dv=e^{\frac{x}{2}} \rightarrow du=4x^3 dx, v=2e^{\frac{x}{2}}$

Apply a partial fraction to the second term in the result of the partial above. Eventually, the partial fraction continues until the x of the term is removed.

```
>>> integrate(x**4*exp(x/2), x)
(2*x**4 - 16*x**3 + 96*x**2 - 384*x + 768)*exp(x/2)
>>> factor(partIntegralS(x**4, exp(x/2), x))
2*(x**4 - 8*x**3 + 48*x**2 - 192*x + 384)*exp(x/2)
```

2.3.2 Integrating Trigonometric Functions

The integral of the trigonometric function is as follows.

$$\int \sin(x)\,dx=-\cos(x)$$
$$\int \cos(x)\,dx=\sin(x)$$
$$\int \tan(x)\,dx=-\ln(\cos(x))$$
$$\int \csc(x)\,dx=\frac{1}{2}(\ln(\cos(x)-1)-\ln(\cos(x)+1))$$
$$\int \sec(x)\,dx=\frac{1}{2}(\ln(\sin(x)-1)-\ln(\sin(x)+1))$$
$$\int \cot(x)\,dx=\ln(\sin(x))$$

These formulas can be derived from the derivative of the trigonometric function as follows:

$(\sin(x))'=\cos(x)$
$\rightarrow \int \cos(x)dx=\sin(x)+c$

$(\cos(x))'=-\sin(x)$
$\rightarrow \int \sin(x)dx=-\cos(x)+c$

$\int \tan(x)dx=\int \dfrac{\sin(x)}{\cos(x)}dx$
$\rightarrow \int -\dfrac{1}{u}du=-\ln(u)+c=-\ln(\cos(x))+c$
$\therefore \cos(x)=u,-\sin(x)dx=du$

Complex trigonometric integrals can be calculated by applying a substitution integral with the above formulas.

Ex) $\int \cos(x)\sin^5(x)\,dx$
$\sin(x)=u,\ \cos(x)dx=du$
$\rightarrow \int u^5\,du = \dfrac{1}{6}u^6+c = \dfrac{1}{6}\sin^6(x)+c$

To apply the substitution integral, the derivative term of the substituted part must be included. In the case of trigonometric functions, various formulas of trigonometric functions are used to make the substitution integral applicable.

Ex) $\int \sin^5(x)\,dx$

$\sin^2(x)+\cos^2(x)=1 \rightarrow \sin^2(x)=1-\cos^2(x)$

Therefore;

$\int \sin^5(x)\,dx = \int \sin(x)\sin^4(x)\,dx = \int \sin(x)(1-\cos^2(x))^2\,dx$

$\cos(x)=u, \rightarrow \sin(x)\,dx=-du$

$-\int (1-u^2)^2\,du = -\int 1-2u^2+u^4\,du$

$$= -u+\frac{2}{3}u^3-\frac{1}{5}u^5+c$$

$$= -\cos(x)+\frac{2}{3}\cos^3(x)-\frac{1}{5}\cos^5(x)+c$$

>>> x, u=symbols("x, u")

>>>integrate(f, x)

-cos(x)**5/5 + 2*cos(x)**3/3 - cos(x)

>>> f=sin(x)*(1-cos(x)**2)**2;f

(-cos(x)**2 + 1)**2*sin(x)

>>> t=cos(x)

>>> dt=diff(t);dt

-sin(x)

>>> fu=f.subs({t:u, sin(x):sin(x)/dt});fu

-(-u**2 + 1)**2

>>> Fu=integrate(fu, u);Fu

-u**5/5 + 2*u**3/3 - u

>>> F=Fu.subs(u, cos(x));F

-cos(x)**5/5 + 2*cos(x)**3/3 - cos(x)

Ex) Integrate the following?

158

$$\int \sin^6(x)\cos^3(x)\,dx$$

If 'cos (x)' in the above function f is replaced with 'u', some of the functions are not replaced.

$$\cos(x) = u, \ -\sin(x)dx = du$$
$$\int \sin^5(x)\sin(x)\cos^3(x)\,dx = -\int \sin^5(x)u^3\,du$$

On the other hand, if 'sin(x)' is replaced by 'u', then all parts of the function are transformed.

$$\sin(x) = u, \ \cos(x)dx = du$$
$$\int \sin^6(x)\cos^2(x)\cos(x)\,dx = \int \sin^6(x)(1-\sin^2(x))\cos(x)\,dx$$
$$= \int u^6(1-u^2)\,du$$
$$= \frac{1}{7}u^7 - \frac{1}{9}u^9 + c$$
$$= \frac{1}{7}\sin^7(x) - \frac{1}{9}\sin^9(x) + c$$

```
>>> x, u=symbols("x, u")
>>> f=sin(x)**6*(1-sin(x)**2)*cos(x);f
(-sin(x)**2 + 1)*sin(x)**6*cos(x)
>>> integrate(f, x)
-sin(x)**9/9 + sin(x)**7/7

>>> t=sin(x)
>>> dt=diff(t);dt
cos(x)
>>> fu=f.subs({t:u, cos(x):cos(x)/dt});fu
u**6*(-u**2 + 1)
>>> Fu=integrate(fu, u);Fu
```

-u**9/9 + u**7/7

>>> F=Fu.subs(u, sin(x));F

-sin(x)**9/9 + sin(x)**7/7

The conversion process of trigonometric functions similar to the above can be summarized as follows.

<Trigonometric substitution method 1>

$$\int \sin^n(x)\cos^m(x)dx$$

If n or m= odd number, it can be transformed into an even number and the remainder can be replaced with a derivative value.

$$\sin^5(x)=\sin^4(x)\sin(x)\rightarrow(1-\cos^2(x))^2\sin(x)$$

As a result, in the trigonometric integral of the above form, the replacement part is set to the odd part of each function's exponent.

A substitution integral can not be applied if the expression does not contain a differentiated term. In this case, the expression can be transformed by the sum, product, or double angle formula of trigonometric functions.

<Trigonometric substitution method 2>

Double angle & Sum formula;

$\cos(2x) = \cos(x+x)$

$\qquad = \cos(x)\cos(x) - \sin(x)\sin(x)$

$\qquad = \cos^2(x) - \sin^2(x)$

$\qquad = 2\cos^2(x) - 1$

$\sin(2x) = \sin(x+x)$

$\qquad = \sin(x)\cos(x) + \cos(x)\sin(x)$

$\qquad = 2\sin(x)\cos(x)$

$\cos^2(x) = \dfrac{1 + \cos(2x)}{2}$

$\sin^2(x) = 1 - \cos^2(x)$

$\qquad = \dfrac{1 - \cos(2x)}{2}$

Product formula;

$\sin(x)\cos(y) = \dfrac{\sin(x-y) - \sin(x+y)}{2}$

$\sin(x)\sin(y) = \dfrac{\cos(x-y) + \cos(x+y)}{2}$

$\cos(x)\cos(y) = \dfrac{\cos(x-y) - \cos(x+y)}{2}$

Ex) Integrate the following?

$$\int \sin^2(x)\cos^2(x)\,dx = \int \frac{1-\cos(2x)}{2} \cdot \frac{1+\cos(2x)}{2}\,dx$$

$$= \frac{1}{4}\int 1-\cos^2(2x)\,dx$$

$$= \frac{1}{4}\int 1-\frac{1}{2}(1+\cos(4x))\,dx$$

$$= \frac{1}{8}x-\frac{1}{8}\int \cos(4x)\,dx$$

$$= \frac{1}{8}x-\frac{1}{32}\sin(4x)+c$$

$$\because 4x=u,\ dx=\frac{1}{4}du$$

$$\int \cos(4x)\,dx = \int \frac{1}{4}\cos(u)\,du = \frac{1}{4}\sin(u)+c = \frac{1}{4}\cos(u)+c$$

In the above results, sin(4x) can be transformed as follows.

sin(4x) = 2sin(2x) cos(2x)

The result of integrate() is the result of applying the transformation of sin(4x).

```
>>> integrate(Rational("1/4")*(1-cos(2*x)**2), x)
x/8 - sin(2*x)*cos(2*x)/16
>>> integrate(sin(x)**2*cos(x)**2, x)
x/8 - sin(2*x)*cos(2*x)/16
```

Ex) Integrate the following?

$$\int \cos(15x)\cos(4x)\,dx$$

The above function is transformed as follows.

$$\cos(15x+4x)=\cos(15x)\cos(4x)-\sin(15x)\sin(4x)$$
$$+\ \underline{\cos(15x-4x)=\cos(15x)\cos(4x)+\sin(15x)\sin(4x)}$$
$$\cos(15x+4x)+\cos(15x-4x)=2\cos(15x)\cos(4x)$$
$$\rightarrow\ \cos(15x)\cos(4x)=\frac{1}{2}(\cos(19x)+\cos(11x))$$

$$\int \cos(15x)\cos(4x)dx$$
$$=\frac{1}{2}\int (\cos(19x)+\cos(11x))dx$$
$$=\frac{1}{2}\left(\frac{1}{19}\sin(19x)+\frac{1}{11}\sin(11x)\right)+c$$

```
>>> re1=integrate(Rational('1/2')*(cos(19*x)+cos(11*x)), x);re1
sin(11*x)/22 + sin(19*x)/38
>>> re2=integrate(cos(15*x)*cos(4*x), x);re2
-4*sin(4*x)*cos(15*x)/209 + 15*sin(15*x)*cos(4*x)/209
```

re1 and re2 must be the same. In order to confirm this, it is judged by substituting an arbitrary value.

```
>>> re1.subs(x, pi/4)
15*sqrt(2)/418
>>> re2.subs(x, pi/4)
15*sqrt(2)/418
```

<Trigonometric substitution method 3>

The expressions related to tan (x) and sec (x) can be transformed by applying the following.

$\sin^2(x)+\cos^2(x)=1$

$$\frac{\sin^2(x)}{\cos^2(x)}+1=\sec^2(x)\rightarrow\tan^2(x)=\sec^2(x)-1$$

$$\sec(x)dx=\frac{1}{\cos(x)}\,dx=\frac{\sin(x)}{\cos^2(x)}=\tan(x)\sec(x)$$

Ex) Integrate the following?

$$\int \sec^9(x)\tan^5(x)dx = \int \sec^8(x)\tan^4(x)\sec(x)\tan(x)\,dx$$

$$= \int \sec^8(x)(\sec^2(x)-1)^2\sec(x)\tan(x)\,dx$$

Substitution: $\sec(x)=u$, $\tan(x)\sec(x)dx=du$

$$\rightarrow \int u^8(u^2-1)^2du = \int u^{12}-2u^{10}+u^8\,du$$

$$= \frac{1}{13}u^{13}-\frac{2}{11}u^{11}-\frac{1}{9}u^9+c$$

$$= \frac{1}{13}\sec^{13}(x)-\frac{2}{11}\sec^{11}(x)+\frac{1}{9}\sec^9(x)+c$$

```
>>> x, u=symbols("x, u")
>>> f=sec(x)**9*tan(x)**5;f
tan(x)**5*sec(x)**9
>>> F1=integrate(f, x);F1
(143*cos(x)**4 - 234*cos(x)**2 + 99)/(1287*cos(x)**13)

>>> f1=sec(x)**8*(sec(x)**2-1)**2*tan(x)*sec(x);f1
(sec(x)**2 - 1)**2*tan(x)*sec(x)**9
>>> t=sec(x)
>>> dt=diff(sec(x));dt
tan(x)*sec(x)
>>> f1u=f1.subs({t:u, tan(x)*sec(x):tan(x)*sec(x)/dt});f1u
u**8*(u**2 - 1)**2
>>> Fu=integrate(f1u, u);Fu
```

```
u**13/13 - 2*u**11/11 + u**9/9
>>> F=Fu.subs(u,sec(x));F
sec(x)**13/13 - 2*sec(x)**11/11 + sec(x)**9/9
```

F1 and F must be the same. Assign π for confirmation.

```
>>> F1.subs(x, pi)
-8/1287
>>> F.subs(x, pi)
-8/1287
```

Ex) Integrate the following.

1) $\int \dfrac{\sqrt{25x^2-4}}{x}\,dx$

If the integrate() function is applied to the above expression, the Piecewise() function is returned. This means that it is a piecewise function whose value depends on the condition of the variable. For the above function, $25x^2-4$ must be greater than or equal to 0.

```
>>> sol=solve(25*x**2-4, x);sol
[-2/5, 2/5]
```

$\sqrt{25x^2-4} \rightarrow 25x^2-4>0$

$$\begin{cases} \sqrt{25x^2-4} \in \text{real number} & \text{if } x \leq -\dfrac{2}{5} \text{ and } x \geq \dfrac{2}{5} \\ \sqrt{25x^2-4} \in \text{complex number} & \text{if } -\dfrac{2}{5} \leq x \leq \dfrac{2}{5} \end{cases}$$

Therefore, the integral result will be a real or complex value depending on the range of x.

```
>>> F=integrate(f, x);F
Piecewise(((-5*I*x/sqrt(-1 + 4/(25*x**2)) -
2*I*Abs(x)*acosh(2/(5*Abs(x)))/x + pi*Abs(x)/x + 4*I/(5*x*sqrt(-1
+ 4/(25*x**2)))), 4/(25*x**2) > 1), (5*x/sqrt(1 - 4/(25*x**2)) +
2*Abs(x)*asin(2/(5*Abs(x)))/x - 4/(5*x*sqrt(1 - 4/(25*x**2)))),
True))
```

The above results are as follows.

$$F=\begin{cases} \dfrac{-5i \cdot x}{\sqrt{-1+\dfrac{4}{25x^2}}} -2i \cdot \cosh^{-1}\dfrac{2}{5|x|}\dfrac{|x|}{x}+\dfrac{\pi|x|}{x}+\dfrac{4i}{5x\sqrt{-1+\dfrac{4}{25x^2}}} & if \ \dfrac{4}{25x^2}>1 \\[20pt] \dfrac{5x}{\sqrt{1-\dfrac{4}{25x^2}}}+sin^{-1}(\dfrac{2}{5|x|})\dfrac{|x|}{x}-\dfrac{4}{5x\sqrt{1-\dfrac{4}{25x^2}}} & if \ \dfrac{4}{25x^2} \leq 1 \end{cases}$$

```
>>> F.subs(x, -3/5)
-3.69552328995372
>>> F.subs(x, -1/5)
-1.0*pi + 0.901864986280756*I
```

It is very difficult to carry out the integration of the above functions manually. Instead, it can be solved more simply by applying a substitution integral.

If the numerator of the above equation is replaced with 'u', there is no differential term of 50xdx=du, so the substitution integral can not be directly applied. In this case, the expression can be converted to a trigonometric function.

Assuming that the numerator of the above formula is high as shown in the triangle below, the hypotenuse and base length are set by Pythagorean theorem.

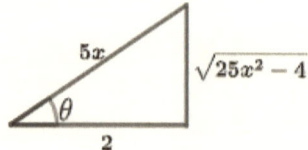

From the above triangle, the following can be defined:

$$\cos(\theta)=\frac{2}{5x}, \quad x=\frac{2}{5}\sec(\theta)\rightarrow dx=\frac{2}{5}\sec(\theta)\tan(\theta)d\theta$$

$$\tan(\theta)=\frac{\sqrt{25x^2-4}}{2} \rightarrow \sqrt{25x^2-4}=2\tan(\theta)$$

Therefore, the above integral can be transformed into a trigonometric integral. In the integral of the following trigonometric function, the derivative of $\tan(\theta)$ is applied for $\sec^2(\theta)$. The derivative of this trigonometric function can be calculated by applying chain rules.

$$\int \frac{2\tan(\theta)}{\frac{2}{5}\sec(\theta)} \cdot \frac{2}{5}\sec(\theta)\tan(\theta)d\theta = 2\int \tan^2(\theta)d\theta$$

$$= 2\int (\sec^2(\theta)-1)d\theta$$

$$= 2\int (\frac{d}{d\theta}\tan(\theta)-1)d\theta$$

$$= 2(\tan(\theta)-\theta)+c$$

$$\because \tan^2(\theta)=\frac{\sin^2(\theta)}{\cos^2(\theta)}=\frac{1-\cos^2(\theta)}{\cos^2(\theta)}=\sec^2(\theta)-1$$

$$\frac{d}{d\theta}\tan(\theta) = \left(\frac{\sin(\theta)}{\cos(\theta)}\right)'$$

$$= \frac{\cos(\theta)\cos(\theta)-\sin(\theta)(-\cos(\theta))}{\cos^2(\theta)}$$

$$= \frac{1}{\cos^2(\theta)}$$

$$= \sec^2(\theta)$$

In above process, tan²(Θ) is transformed as the following.

$$x=\frac{2}{5}\sec(\theta) \rightarrow \theta=\sec^{-1}(\frac{5}{2}x)=\text{asec}(\frac{5}{2}x)$$

```
>>> th=symbols("th")
>>> F1=integrate(2*tan(th)**2, th);F1
-2*th + 2*sin(th)/cos(th)
```

As a result of the two integrals, the shapes of F and F1 are different, but the definite integral for interval [1,5] is the same.

```
>>> N(F.subs(x, 5)-F.subs(x, 1))
19.6744333617307
```

The variable in the above function F1 is θ. Since the above interval is for variable x, F1 must be converted to variable x. Use the inverse trigonometric function for this conversion.

$$x=\frac{2}{5}\sec(\theta) \rightarrow \theta=\sec^{-1}(\frac{5}{2}x)=\text{asec}(\frac{5}{2}x)$$

```
>>> F2=F1.subs(th, asec(5*x/2));F2
2*sin(asec(5*x/2))/cos(asec(5*x/2)) - 2*asec(5*x/2)
>>> N(F2.subs(x, 5)-F2.subs(x,1))
19.6744333617307
```

2) $\int_{\frac{-4}{5}}^{\frac{-2}{5}} \frac{\sqrt{25x^2-4}}{x}\, dx$

The variable of the integral shown in problem 1) is theta, but the upper and lower bounds of the definite integral are x units. Therefore, their units should be converted to theta.

$$\cos(\theta)=\frac{2}{5x} \rightarrow \begin{cases} x=\frac{-2}{5} \rightarrow \cos(\theta)=\frac{2}{5}\frac{-5}{2}=-1 \rightarrow \theta=\cos^{-1}(-1) \\ x=\frac{-4}{5} \rightarrow \cos(\theta)=\frac{2}{5}\frac{-5}{4}=-\frac{1}{2} \rightarrow \theta=\cos^{-1}(-\frac{1}{2}) \end{cases}$$

```
>>> theta=acos(2/(5*x));theta
acos(2/(5*x))
>>> up=theta.subs(x, -2/5);up
3.14159265358979
>>> low=theta.subs(x, -4/5);low
2.09439510239320
>>> val=integrate(f1, (th, low, up));val
```

1.36970651274456

The upper limit of the integral, -2/5, violates the condition of $25x^2-4>0$ (nominator). Therefore, the definite integral can not be calculated. Instead, if the definite integral is calculated as a value close to the upper limit, it is similar to the above result.

```
>>> N(F.subs(x, Rational(-2.00000001, 5))-F.subs(x, Rational(-4, 5)),
4)
1.370
```

Ex) Integrate the following.

$$\int \frac{1}{x^2+a^2} \, dx$$

$x^2+a^2 \neq 0$ at the above equation. Therefore, when defining variables with symbols(), use the positive argument.

```
>>> x, a=symbols("x,a", positive=True)
>>> f=1/(x**2+a**2);f
1/(a**2 + x**2)
>>> F=integrate(f, x);F
atan(x/a)/a
```

The above procedure can be understood by applying a trigonometric function. As shown in the following Figure, the denominator of the above equation can be regarded as the length of the hypotenuse along the length x and a of the side.

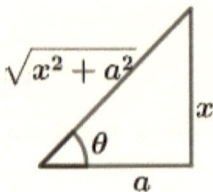

From the above Figure,

$$\tan(\theta) = \frac{x}{a} \rightarrow x = a \cdot \tan(\theta), \quad \theta = \tan^{-1}\left(\frac{x}{a}\right)$$

$$\frac{dx}{d\theta} = a\left(\frac{\sin(\theta)}{\cos(\theta)}\right)' = a\left(\frac{\cos(\theta)\cos(\theta) + \sin(\theta)\sin(\theta)}{\cos^2(\theta)}\right) = a \cdot \sec^2(\theta)$$

$$\frac{1}{x^2 + a^2} = \frac{1}{a^2(\tan^2(\theta) + 1)} = \frac{1}{a^2\sec^2(\theta)}$$

$$\because \tan^2(\theta) + 1 = \frac{\sin^2(\theta)}{\cos^2(\theta)} + \frac{\cos^2(\theta)}{\cos^2(\theta)} = \frac{1}{\cos^2(\theta)} = \sec^2(\theta)$$

$$\rightarrow \int \frac{1}{a^2 \cdot \sec^2(\theta)} a \cdot \sec^2(\theta) d\theta$$

$$= \frac{1}{a} \int d\theta$$

$$= \frac{1}{a} \theta + c$$

$$= \frac{1}{a} \tan^{-1}\left(\frac{x}{a}\right) + c$$

Ex) Integrate the following.

$$\int \sqrt{x^2 + 4x + 5} \ dx$$

Direct integration of the above functions is not done.

```
>>> x, a=symbols("x,a", real=True)
>>> f=sqrt(x**2+4*x+5);f
```

```
sqrt(x**2 + 4*x + 5)
>>> F=integrate(f, x);F
Integral(sqrt(x**2 + 4*x + 5), x)
```

Therefore, the above function can also be calculated by switching to a trigonometric function.

$$x^2+4x+5=(x+2)^2+1$$

From the above equation, the following triangles can be created.

$$A(x)=\pi f^2(x)-\pi g^2(x)=pi\left(x^{\frac{2}{3}}-\left(\frac{x}{4}\right)^2\right)$$

$$x^{\frac{1}{3}}-\frac{x}{4}=0 \rightarrow x=0,\ 8$$

$$\int_0^8 A(x)\,dx=\pi\int_0^8\left(x^{\frac{2}{3}}-\left(\frac{x}{4}\right)^2\right)dx=\frac{128}{5}\pi$$

From the triangle shown in the above example,

$$\tan(\theta)=x+2 \rightarrow \sec^2(\theta)d\theta=dx$$

$$\int \sqrt{(x+2)^2+1}\ dx=\int \sqrt{\tan^2(\theta)+1}\ \sec^2(\theta)d\theta=\int \sec^3(\theta)\,d\theta$$

$$\because \tan^2(\theta)+1=\frac{\sin^2(\theta)+\cos^2(\theta)}{\cos^2(\theta)}=\frac{1}{\cos^2(\theta)}=\sec^2(\theta)$$

Applying partial integral for $\sec^3(\theta)=\sec^2(\theta)\sec(\theta)$

$$\int \sec^3(\theta)\,d\theta = \int \sec(\theta)\sec^2(\theta)\,d\theta$$

$$= \sec(\theta)\tan(\theta) - \int \tan(\theta)\sec(\theta)\tan(\theta)\,d\theta$$

$$= \sec(\theta)\tan(\theta) - \int (\sec^2(\theta)-1)\sec(\theta)\,d\theta$$

$$= \sec(\theta)\tan(\theta) - \int \sec^3(\theta)\,d\theta + \int \sec(\theta)\,d\theta$$

$$\because (\sec(\theta))\;' = \left(\frac{1}{\cos(\theta)}\right)'\;' = \frac{\sin(\theta)}{\cos^2(\theta)}\;' = \tan(\theta)\sec(\theta)$$

$$\int \sec^2(\theta)\,d\theta\;' = \int \frac{d}{d\theta}\tan(\theta)\,d\theta\;' = \tan(\theta)$$

$$\int \sec^3(\theta)\,d\theta = \frac{1}{2}\sec(\theta)\tan(\theta) - \frac{1}{4}(\ln(\sin(\theta)-1)+\ln(\sin(\theta)+1))+c$$

$$= \frac{1}{2}(x+2)\sqrt{x^2+4x+5} - \frac{1}{4}(\ln(\sqrt{x^2+4x+5}-1)+\ln(\sqrt{x^2+4x+5}+1))+c$$

$$\because \int \sec(x)\,dx = \frac{1}{2}(\ln(\sin(x)-1)-\ln(\sin(x)+1))$$

>>> F=integrate(sec(t)**3, t);F

-log(sin(t) - 1)/4 + log(sin(t) + 1)/4 - sin(t)/(2*sin(t)**2 - 2)

In the result,

$$\frac{\sin(t)}{\sin^2(t)-1} = \frac{\sin(t)}{\cos^2(t)} = \frac{\sin(t)}{\cos(t)}\frac{1}{\cos(t)} = \tan(t)\sec(t)$$

2.3.3 Partial fraction decomposition

In the case of integrating a polynomial of a fractional form, a substitution integral can be applied if the derivative of the denominator is the same as the numerator.

$$\int \frac{2x-1}{x^2-x-6}\,dx = \int \frac{1}{u}\,du = \ln(u)+c = \ln(x^2-x-6)+c$$

$$\because x^2-x-6=u \rightarrow (2x-1)\,dx = du$$

Unlike the above example, it is difficult to calculate the integral for complex fractions. Therefore, it needs to be converted into a simple form. Such a transformation can be achieved by partial fraction decomposition.

<Partial Fraction>

This is a method of factoring the denominator and then generating new fractions with the denominator of each argument.

$$f(x) = \frac{p(x)}{q(x)}$$

p(x), q (x) are coprime, ie they have no common argument. The order of the numerator is smaller than the order of the denominator.

Partial fractions can be divided by the factorization of the denominator, and the numerator is set lower by one order than each factorized factor.

For example,

$$\frac{2x-22}{x^2-x+12} = \frac{a}{x+3} + \frac{b}{x-4} = \frac{ax-4a+bx+3b}{(x+3)(x-4)}$$

$\because$ denominator: $x^2-x+12 = (x+3)(x-4)$

$2x-22 = (a+b)x-4a+3b \rightarrow a+b = 2, -4a+3b = -22$

$a = 4, b = -2$

$$\Rightarrow \frac{2x-22}{x^2-x+12} = \frac{4}{x+3} - \frac{2}{x-4}$$

These partial fractions are calculated using the apart() function of the sympy function.

```
>>> f=(2*x-22)/(x**2-x-12);f
(2*x - 22)/(x**2 - x - 12)
>>> f1=apart(f);f1
4/(x + 3) - 2/(x - 4)
```

Ex) Integrate the following.

$$\int \frac{x^2+4}{3x^3+4x^2-4x} \, dx$$

Factorization of denominator

```
>>> factor(3*x**3+4*x**2-4*x)
x*(x + 2)*(3*x - 2)
>>> f=a/x+b/(x+2)+c/(3*x-2)
```

$$\frac{x^2+4}{3x^3+4x^2-4x} = \frac{a}{x} + \frac{b}{x+2} + \frac{c}{3x-2}$$

a, b, c?

The above equation is used to compare the numerators of the left and right terms. Reduction to common denominator in the right-hand side uses the together() function. The fractions generated by this function can be represented by separating the numerator and the denominator using the numer() and denom() functions, respectively.

```
>>> f1=together(f);f1  #reduction to common denominator
```

$(a*(x + 2)*(3*x - 2) + b*x*(3*x - 2) + c*x*(x + 2))/(x*(x + 2)*(3*x - 2))$

>>> exp=expand(numer(f1));exp #express numerator

$3*a*x**2 + 4*a*x - 4*a + 3*b*x**2 - 2*b*x + c*x**2 + 2*c*x$

>>> denom(f1) #express denomintor

$x*(x + 2)*(3*x - 2)$

>>> exp1=collect(exp,x);exp1

$-4*a + x**2*(3*a + 3*b + c) + x*(4*a - 2*b + 2*c)$

>>> eq1=exp1.coeff(x, 2)-1;eq1

$3*a + 3*b + c - 1$

>>>eq2=exp1.coeff(x, 1);eq

$3*a*x**2 + 4*a*x - 4*a + 3*b*x**2 - 2*b*x + c*x**2 + 2*c*x$

>>>eq3=exp1.coeff(x, 0)-4

$-4*a - 4$

>>>solve([eq1,eq2,eq3],(a,b,c))

{a: -1, b: 1/2, c: 5/2}

$$2x-1=x^2(3a+3b+c)+x(4a-2b+2c)-4a$$
$$2=3a+3b+c, \quad 0=4a-2b+2c, \quad -1=-4a$$
$$\rightarrow \quad 3a+3b+c=1, \quad 4a-2b+2c=0, \quad -4a=4$$
$$\rightarrow \quad a=-1, \quad b=\frac{1}{2}, \quad c=\frac{5}{2}$$

>>> f.subs({a:sol[a], b:sol[b], c:sol[c]})

$5/(2*(3*x - 2)) + 1/(2*(x + 2)) - 1/x$

$$\frac{-1}{x} + \frac{1}{2(x+2)} + \frac{5}{2(3x-2)}$$

The above procedure is performed identically by the apart() function.

>>> x=symbols("x")

176

```
>>> f=(x**2+4)/(3*x**3+4*x**2-4*x);f
(x**2 + 4)/(3*x**3 + 4*x**2 - 4*x)
>>> f1=apart(f);f1
5/(2*(3*x - 2)) + 1/(2*(x + 2)) - 1/x
```

Integrate the partial fraction

$$\int \frac{-1}{x}+\frac{1}{2(x+2)}+\frac{5}{2(3x-2)}\,dx = -\int \frac{1}{x}\,dx+\frac{1}{2}\int \frac{1}{x+2}\,dx+\frac{5}{2}\int \frac{1}{3x-2}\,dx$$

$$= -\ln(x)+\frac{1}{2}\ln(x+2)+\frac{5}{6}\ln(3x-2)+c$$

← 3x-2=u, 3dx=du

$$\int \frac{1}{3x-2}\,dx=\frac{1}{3}\int u\,du=\frac{1}{3}\ln(u)+c=\frac{1}{3}\ln(3x-2)+c$$

```
>>> integrate(f1, x)
-log(x) + 5*log(x - 2/3)/6 + log(x + 2)/2
```

It is equivalent to dividing the original function directly.

```
>>> integrate(f, x)
-log(x) + 5*log(x - 2/3)/6 + log(x + 2)/2
```

Ex) Partial fractional decomposition and integration of the following functions.

$$f(x)=\frac{x^3+10x^2+3x+36}{(x-1)(x^2+4)^2}$$

Partial fractions are generated by the apart() function and are:

```
>>> x, A, B, C,D,E=symbols("x, A, B, C,D,E", real=True)
>>>f=(x**3+10*x**2+3*x+36)/((x-1)*(x**2+4)**2);f
(x**3 + 10*x**2 + 3*x + 36)/((x - 1)*(x**2 + 4)**2)
>>>f1=apart(f);f1
```

x/(x**2 + 4)**2 - (2*x + 1)/(x**2 + 4) + 2/(x - 1)

The above process is as follows.

1) The denominator of the above function can be factorized into (x-1), (x^2 + 4), and (x^2 + 4)^2.

2) The numerator of each partial fraction should be set one dimension lower than the denominator.

In the case of (x^2 + 4)^2 the molecule should be set based on the irreducible denominator (x^2 + 4).

Therefore, the partial fraction has the form:

$$\frac{x^3+10x^2+3x+36}{(x-1)(x^2+4)^2} = \frac{A}{x-1} + \frac{Bx+C}{x^2+4} + \frac{Dx+E}{(x^2+4)^2}$$

$$= \frac{A(x^2+4)^2+(x-1)(x^2+4)(Bx+C)+(x-1)(Dx+E)}{(x-1)(x^2+4)^2}$$

$$= \frac{16A-4C-E+x^4(A+B)+x^3(-B+C)+x^2(8A+4B-C+D)+x(-4B+4C-D+E)}{(x-1)(x^2+4)^2}$$

>>> fp=A/(x-1)+(B*x+C)/(x**2+4)+(D*x+E)/(x**2+4)**2;fp
A/(x - 1) + (B*x + C)/(x**2 + 4) + (D*x + E)/(x**2 + 4)**2
>>> fpt=together(fp);fpt
(A*(x**2 + 4)**2 + (x - 1)*(x**2 + 4)*(B*x + C) + (x - 1)*(D*x + E))/((x - 1)*(x**2 + 4)**2)

Each coefficient A, B, C, D, and E can be calculated by comparing the right-hand and left-hand terms of the above equation. Only numerators of the above formulas are considered for this comparison.

>>> eq=Eq(numer(f), numer(fpt));eq

$$Eq(x^{**}3 + 10^*x^{**}2 + 3^*x + 36, A^*(x^{**}2 + 4)^{**}2 + (x - 1)^*(x^{**}2 + 4)^*(B^*x + C) + (x - 1)^*(D^*x + E))$$

<u>solve undetermined coeffs()</u> function is applied to calculate the coefficients in above equation. This function is a function for calculating multiple coefficients in one expression rather than a simultaneous equation.

```
>>> coef=solve_undetermined_coeffs(eq, [A,B,C,D,E], x);coef
{A: 2, B: -2, C: -1, D: 1, E: 0}
>>> fp.subs(coef)
x/(x**2 + 4)**2 + (-2*x - 1)/(x**2 + 4) + 2/(x - 1)
```

Finally, the partial fractions are:

$$\frac{x^3+10x^2+3x+36}{(x-1)(x^2+4)^2} = \frac{2}{x-1} - \frac{2x+1}{x^2+4} + \frac{x}{(x^2+4)^2}$$

The integral of this function is as follows.(<u>As a result of integration, refer to the example above for atan.</u>)

$$\int \frac{2}{x-1} - \frac{2x+1}{x^2+4} + \frac{x}{(x^2+4)^2}\, dx = \int \frac{2}{x-1}\, dx - \int \frac{2x+1}{x^2+4}\, dx + \int \frac{x}{(x^2+4)^2}\, dx$$

$$= 2\ln(x-1)+\ln(x^2+4)+\tan^{-1}\left(\frac{x}{2}\right)-\frac{1}{2}\frac{1}{x^2+4}+c$$

$$\int \frac{2}{x-1}\, dx = 2\ln(x-1)+c$$

$$\int \frac{2x+1}{x^2+4}\, dx = \int \frac{2x}{x^2+4}\, dx + \int \frac{1}{x^2+4}\, dx = \ln(x^2+4)+\tan^{-1}(\frac{x}{2})+c$$

$$x^2+4=u,\ 2xdx=du \longrightarrow \int \frac{2x}{x^2+4}\, dx = \int \frac{1}{u}\, du = \ln(u)+c = \ln(x^2+4)+c$$

$$\frac{x}{2}=v,\ \frac{1}{2}dx=v \longrightarrow \int \frac{1}{x^2+4}\, dx = \frac{1}{2}\int \frac{\frac{1}{2}}{\left(\frac{x}{2}\right)^2+1}\, dx = \frac{1}{2}\int \frac{1}{v^2+1}\, dv = \tan^{-1}\left(\frac{x}{2}\right)+c$$

```
>>> integrate(f1, x)
2*log(x - 1) - log(x**2 + 4) - atan(x/2)/2 - 1/(2*x**2 + 8)
```

The above result is the same as the integral for the original function.

```
>>> integrate(f)
2*log(x - 1) - log(x**2 + 4) - atan(x/2)/2 - 1/(2*x**2 + 8)
```

Ex) Integrate the following.

$$\int \frac{x^2}{x^2-1} dx = \int 1 - \frac{1}{2(x+1)} + \frac{1}{2(x-1)} dx$$

$$= x - \frac{1}{2}\ln(x-1) + \frac{1}{2}\ln(x+1) + c$$

```
>>> x=symbols("x")
>>> f=x**2/(x**2-1);f
x**2/(x**2 - 1)
>>> f1=apart(f); f1
1 - 1/(2*(x + 1)) + 1/(2*(x - 1))
>>> integrate(f1, x)
x + log(x - 1)/2 - log(x + 1)/2
>>> integrate(f, x)
x + log(x - 1)/2 - log(x + 1)/2
```

2.3.4 Integration of functions with roots

In general, the integral of a function containing root applies the substitution method as follows.

$$\sqrt[n]{g(x)} = u \rightarrow g(x) = u^n \rightarrow g(x)dx = nu^{n-1}du$$

In general, the substitution integral can be applied when the differential term of the substituted part is included in the equation. The substitution of the root term can be applied regardless of the existence of the differential term.

Ex) Integrate the follow.

$$\int \frac{x+2}{\sqrt[3]{x-3}} dx$$

$$\sqrt[3]{x-3} = u, \ x = u^3 + 3 \rightarrow dx = 3u^2 du$$

$$\int \frac{u^3+5}{u} 3u^2 du = \int 3u^4 + 15u \, du$$

$$= \frac{3}{5}u^5 + \frac{15}{2}u^2 + c$$

$$= \frac{3}{5}\sqrt[3]{(x-3)^5} + \frac{15}{2}\sqrt[3]{(x-3)^2} + c$$

If the above function is a direct integral function, a piecewise function is created because of the fractional condition (denominator≠0).

```
>>> x, u=symbols("x, u", real=True)
>>> f=(x+2)/(x-3)**Rational(1,3);f
(x + 2)/(x - 3)**(1/3)
>>>Fd=integrate(f, x);Fd
```

Piecewise((3*(x - 3)**(2/3)*(x + 2)/5 + 9*(x - 3)**(2/3)/2, Abs(x + 2)/5 > 1), (-3*(-x + 3)**(2/3)*(x + 2)*exp(5*I*pi/3)/5 - 9*(-x + 3)**(2/3)*exp(5*I*pi/3)/2, True))

The above piecewise function can be applied to all ranges of the variable. It is accurate, but it is difficult to calculate manually. Instead, it can be computed more easily by applying substitution integral.

```
>>> u=symbols('u')
>>> a=solve(((x-3)**Rational("1/3")-u),x);a
[u**3 + 3]
```

The solve() expression evaluates the solution of the specified variable.

$$u=\sqrt[3]{x-3} \quad \rightarrow \quad x=u^3+3$$

```
>>> du=diff(sol[0], u);du
3*u**2
>>> f1=f.subs({(x-3)**Rational("1/3"):u,x:sol[0]})*du ;f1
3*u*(u**3 + 5)
>>> Fu=integrate(f1, u);Fu
3*u**5/5 + 15*u**2/2
>>> F=Fu.subs(u,(x-3)**Rational(1,3));F
3*(x - 3)**(5/3)/5 + 15*(x - 3)**(2/3)/2
```

Let's calculate the definite integral between [4, 10] to see if Fd and F are equal.

```
>>> N(Fd.subs(x, 10)-Fd.subs(x, 4),3)
34.7
>>> N(F.subs(x, 10)-F.subs(x, 4), 3)
34.7
```

Ex) Integrate the follow.

$$\int \frac{2}{x-3\sqrt{x+10}}\, dx$$

The above function is a piecewise function.

$$\int \frac{2}{x-3\sqrt{x+10}}\, dx = \begin{cases} \text{exist in real number} & \text{if } x \geq -10 \text{ and } x \neq 15 \\ \text{not exist} & \text{if } x = 15 \\ \text{exist in complex} & \text{if } x \leq -10 \end{cases}$$

Therefore, integrating this function directly returns an integral object, not an integral result. In addition, only the definite integral in the interval where the integral exists, that is, the continuous interval, is calculated.

```
>>> f=2/(x-3*sqrt(x+10));f
2/(x - 3*sqrt(x + 10))
>>> integrate(f, x)
2*Integral(1/(x - 3*sqrt(x + 10)), x)
```

Definite integrating at [1,5]

```
>>> N(integrate(f, (x, 1, 5)).doit())
-1.03262053876493
```

Definite integrating at [-20, -10]

```
>>> N(integrate(f, (x, -20, -10)).doit())
```

183

-1.19653917717051 + 0.460574763391205*I

However, the above problem can be solved by applying the substitution method.

$$\sqrt{x+10} = u, \ x = u^2 - 10 \rightarrow dx = 2u\,du$$

$$\int \frac{2}{x-3\sqrt{x+10}}\,dx = \int \frac{4u}{u^2-10-3u}\,du$$

$$= \int \frac{8}{7(u+2)} + \frac{20}{7(u-5)}\,du$$

$$= \frac{8}{7}\ln(u+2) + \frac{20}{7}\ln(u-5) + c$$

$$= \frac{8}{7}\ln(\sqrt{x+10}+2) + \frac{20}{7}\ln(\sqrt{x+10}-5) + c$$

```
>>> x, u=symbols("x, u", real=True)
f=2/(x-3*sqrt(x+10));f
>>> 2/(x - 3*sqrt(x + 10))
>>> a=solve(sqrt(x+10)-u, x);a
[u**2 - 10]
>>> du=diff(a[0],u);du
2*u
>>> fu=f.subs({sqrt(x+10):u, x:a[0]})*du;fu
4*u/(u**2 - 3*u - 10)
>>> Fu=integrate(fu, u);Fu
20*log(u - 5)/7 + 8*log(u + 2)/7
>>> F=Fu.subs(u, sqrt(x+10));F
20*log(sqrt(x + 10) - 5)/7 + 8*log(sqrt(x + 10) + 2)/7
```

In [1,5], the direct result of the above function must match the value by substitution integral.

```
>>> N(integrate(f, (x, 1, 5)).doit(), 4)
-1.033
>>> N(F.subs(x, 5)-F.subs(x, 1), 4)
-1.033
```

2.4 Application of integral

2.4.1 Mean value theorem

The mean value of f(x) between [a, b] is calculated as follows:

$$f_{avg} = \frac{1}{b-a} \int_a^b f(x)\,dx$$

Ex) Mean value of f(x) in specific interval?

1) $f(t) = t^2 - 5t + 6\cos(\pi t)$, [-1, 5/2]

$$\int_{-1}^{\frac{5}{2}} (t^2 - 5t + 6\cos(\pi t))\,dt = \frac{1}{3}t^3 - \frac{5}{2}t^2 + \frac{6}{\pi}\sin(\pi t)\Big|_{-1}^{\frac{5}{2}}$$

```
>>> t=symbols("t")
>>> MeanValue=integrate(t**2-5*t+6*cos(pi*t), (t, -1,
Rational('5/2')))/(Rational('5/2')--1);MeanValue
-13/6 + 12/(7*pi)
>>> MeanValue.evalf(3)
-1.62
```

x.evalf(): Evaluates the object and returns a value.

2) $R(x) = \sin(2x) \cdot \exp(1 - \cos(2x))$, [-π, π]

By $1 - \cos(2x) = u$, [-π, π]→[0, 0]

Since the upper and lower limits of the integral are the same, the integral of the above function is 0.

186

$$\int_{-\pi}^{\pi} \sin(2x)e^{1-\cos(2x)}dx = \int_{0}^{0} \frac{1}{2}e^{u}du = 0$$

∵ 1-cos(2x)=u, 2sin(2x)dx=du

x=pi→u=0, x=-pi→u=0

```
>>> x, u=symbols('x, u')
>>> R=sin(2*x)*exp(1-cos(2*x));R
 exp(-cos(2*x) + 1)*sin(2*x)
>>> y=1-cos(2*x)
>>> dy=diff(y, x);dy
 2*sin(2*x)
>>>R1=R.subs({y:u, sin(2*x):sin(2*x)/dy});R1
exp(u)/2
>>> R1_int=integrate(R1, (u, 0, 0));R1_int
0
```

Ex) A point c corresponding to the mean value of f(x) in the interval [1, 4]?

$f(x)=x^2+3x+2$

$$f_{avg}=\frac{1}{4-1}\int_{1}^{4} (x^2+3x+2)dx$$

$$= \frac{1}{3}\left(\frac{1}{3}x^3+\frac{3}{2}x^2+2x\right)\Big|_{1}^{4}$$

$$= \frac{33}{2}$$

```
>>> x, c=symbols('x, c')
>>> f=x**2+3*x+2;f
 x**2 + 3*x + 2
>>> f_avg=integrate(f,(x, 1,4))/(4-1);f_avg
```

That is, the following equation is established at a point c of f(x) in [1, 4].

$f(c)=f_{avg}$

Therefore, $c^2+3c+2=33/2$. The equation can be made using Eq() and calculated with solve().

```
>>> eq=Eq(f.subs(x,c), f_avg);eq
Eq(c**2 + 3*c + 2, 33/2)
>>> c_sol=solve(eq, c);c_sol
[-3/2 + sqrt(67)/2, -sqrt(67)/2 - 3/2]
>>> [i.evalf(4) for i in c_sol]
[2.593, -5.593]
```

2.4.2 Applied to curves

Length of a curve

Figure 2.3 draws 10 points on the function to calculate the length of the graph in the interval [a, b]. The length of this curve will be close to the sum of the lengths of the subdivisions of the straight line in that section.

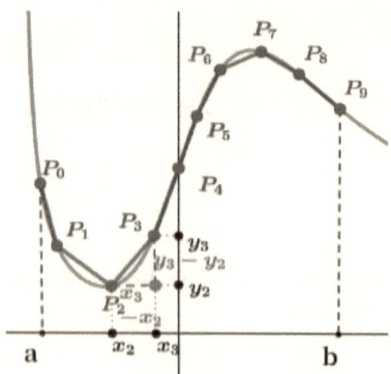

Figure 2.3 The length of curve.

The length of the line P_2P_3 in Figure 2.3 can be calculated by applying Pythagorean theorem.

$$\overline{P_2P_3} = \sqrt{(x_3-x_2)^2+(y_3-y_2)^2}$$

Adding the lengths of all the parts in Figure 2.3 above will approximate the length L of the curve of [a, b].

$$L \approx \sum_{i=1}^{10} \sqrt{(x_i-x_{i-1})^2+(y_i-y_{i-1})^2} = \sum_{i=1}^{10} \sqrt{\Delta x_i^2+\Delta y_i^2}$$

As the number of subdivisions increases, the difference between the length of the curve and the sum of each straight line length will decrease. Therefore, the above equation can be modified as follows.

$$L = \lim_{n \to \infty} \sum_{i=1}^{n} \sqrt{(x_i-x_{i-1})^2+(y_i-y_{i-1})^2} \quad (1)$$

189

f(x_i^*) corresponding to a point x_i^* existing between two points x_i and x_{i+1} in consecutive interval of Equation (1) can be expressed as Equation (2) by applying the <u>intermediate value theorem</u>.

$$f'(x_i^*)=\frac{f(x_i)-f(x_{i-1})}{x_i-x_{i-1}}=\frac{dy}{dx} \qquad (2)$$

<length by curve>

Equation(1) is substituted into equation(2), which is an intermediate value theorem, and the sum of these values can be calculated using the definite integral.

$$L= \lim_{n\to\infty} \sum_{i=1}^{n} \sqrt{(x_i-x_{i-1})^2+(y_i-y_{i-1})^2}$$

$$= \lim_{n\to\infty} \sum_{i=1}^{n} \sqrt{1+\frac{(y_i-y_{i-1})^2}{(x_i-x_{i-1})^2}}\,(x_i-x_{i-1})$$

$$= \int_a^b \sqrt{1+\left(\frac{dy}{dx}\right)^2}\,dx$$

Ex) Find the length of the curve.

1) The length of y=ln(sec(x)) in 0≤x≤π/4?

$$L = \int_0^{\frac{\pi}{4}} \sqrt{1 + \left(\frac{d}{dx}\ln(\sec(x))\right)^2}\ dx$$

$$= \int_0^{\frac{\pi}{4}} \sec(x)\,dx$$

$$= \frac{1}{2}\left(\ln((\sin(x)-1))+\ln(\sin(x)+1)\right)\Big|_0^{\frac{\pi}{4}}$$

$$= 0.8814$$

$$\leftarrow \quad \frac{d}{dx}\ln(\sec(x)) = \tan(x)$$

$$1 + \tan^2(x) = \sec^2(x)$$

```
>>> x=symbols("x")
>>> f=log(sec(x));f
log(sec(x))
>>> diff(f)
tan(x)
>>> integrate(sqrt(1+diff(f)**2), (x, 0, pi/4)).evalf(4)
0.8814
```

2) The length of the following in $1 \le y \le 4$?

$$f(y) = \frac{2}{3}(y-1)^{\frac{3}{2}}$$

$$f'(y) = \frac{2}{3}\frac{3}{2}\sqrt{y-1}$$

$$\int_1^4 \sqrt{1 + \left(\frac{d}{dy}f(y)\right)^2}\ dy = \int_1^4 \sqrt{y}\ dy = \frac{14}{3}$$

```
>>> y=symbols("y", real=True)
>>> f=Rational(2,3)*(y-1)**Rational(3,2);f
2*(y - 1)**(3/2)/3
>>> dy=diff(f, y); dy
```

sqrt(y - 1)

```
>>> L=integrate(sqrt(1+dy**2), (y, 1, 4));L
14/3
```

Area by curves

Ex) Calculate the area of the area shared by $y=x^4$ and $y=x^{0.5}$.

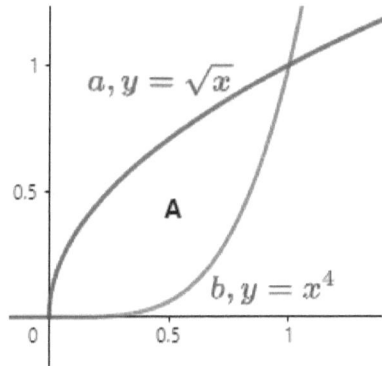

Figure 2.4 Area between two curves.

In Figure 2.4, area A shared by curve a and curve b is calculated as follows.

area of curve a - area of curve b=A

The contact of the two curves is calculated by solve().

$x^{1/2}=x^4$

```
>>> x=symbols('x', real=True)
>>> a=sqrt(x);a
sqrt(x)
>>> b=x**4;b
```

192

x**4

```
>>> eq=Eq(a, b);eq
Eq(sqrt(x), x**4)
>>> sol=solve(eq, x);sol
[0, 1]
```

Both functions intersect at x=0 and 1. Therefore, the difference between the two functions is definite integration at [0, 1].

```
>>> integrate(a-b, (x, 0, 1))
7/15
```

Ex) Ar shown as Figure 2.5?

a(x)=x·exp(-x²), b(x)=x+1, c(x)=2

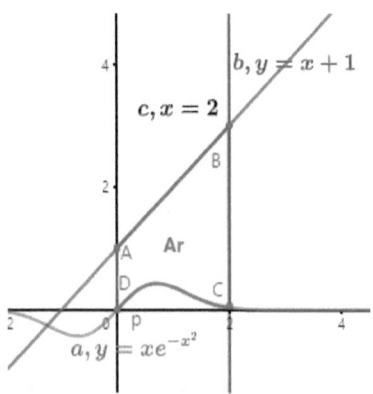

Figure 2.5 Area by three equations.

The integration of b-a in [0,2] =Ar

```
>>> a=x*exp(-x**2);a
x*exp(-x**2)
```

```
>>> b=x+1;b
```
x + 1
```
>>> integrate(b-a, (x, 0, 2)).evalf(4)
```
3.509

Ex) Calculate the shared area by the following functions.

$a(x)=2x^2+10, \ b(x)=4x+16$

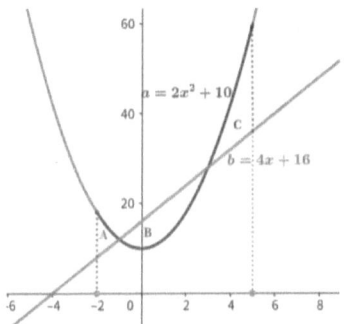

Figure 2.6 Area shared by curve and line.

As shown in Figure 2.6, the area shared by a(x) and b(x) in [-2, 5] should be considered as A, B, and C. Areas are separated by the point where they meet.

```
>>> a=2*x**2+10;a
```
2*x**2 + 10
```
>>> b=4*x+16;b
```
4*x + 16
```
>>> meetP=solvc(a-b, x); meetP
```
[-1, 3]

The subdivisions that determine each area based on the intersection x = -1 and 3 can be divided into [-2, -1], [-1, 3], and [3, 5].

If the shape of the graph is not known, the shared area can be calculated as the absolute value of the difference of the definite integral by each function.

$$\left|\int_{down}^{up} a(x)dx - \int_{down}^{up} b(x)dx\right| = \left|\int_{down}^{up} 2x^2+10\,dx - \int_{down}^{up} 4x+16\,dx\right|$$

```
>>> rng=[[-2, -1], [-1, 3],[3, 5]]
>>> A=[]
>>> for i in rng:
        a1=integrate(a, (x, i[0],i[1]))
        b1=integrate(b, (x, i[0],i[1]))
        A.append(abs(a1-b1))
>>> A
[14/3, 64/3, 64/3]
>>> sum(A)
142/3
```

Surface area and volume of rotating body

If a certain section of the function f(x) is rotated about the x axis, a three-dimensional rotation body appears as shown in Figure 2.7. The surface area of this rotating body can be expressed by integrating the area of a certain section, and the volume can be calculated by integrating the surface area of the rotating body.

195

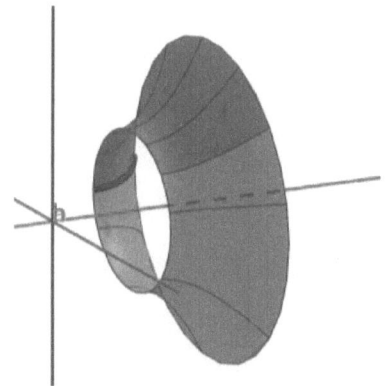

Figure 2.7 Rotation of function f(x).

The surface area A

A=circumference×length = 2πrL

$$r=\frac{1}{2}(r_1+r_2)$$

r_1: radius of the circle at the left end of the section = $f(x_1)$

r_2: radius of the circle at the right end of the section =$f(x_2)$

Since L is the <u>length</u> of the curve, the surface area of the rotating body can be calculated by integrating the cross-sectional area.

$$A=\int_a^b 2\pi f(x)\sqrt{1+(f'(x))^2}\ dx$$

Ex) Find the surface area of the rotating body of the next function.

1) $y=\sqrt{9-x^2}$, $-2\leq x\leq 2$

>>> f=sqrt(9-x**2);f

196

sqrt(-x**2 + 9)

>>> df=diff(f, x);df

-x/sqrt(-x**2 + 9)

>>> A=integrate(2*pi*f*sqrt(1+df**2), (x, -2, 2))

>>> nsimplify(A)

24*pi

2) $y=\sqrt[3]{x}$, $1\leq y\leq 2$
→ $f(y)=y^3$

>>> f=y**3;f

y**3

>>> df=diff(f, y)

>>> A=integrate(2*pi*f*sqrt(1+df**2),(y, 1, 2))

>>> A.evalf(4)

199.5

Ex) Calculate the volume of the rotating body when rotating the function f(x) with respect to the x axis.

In the interval [1, 4], rotating the function f(x) = x²-4x+5 (Figure 2.8a) with respect to the x axis produces a rotation as shown in Figure 2.8b.

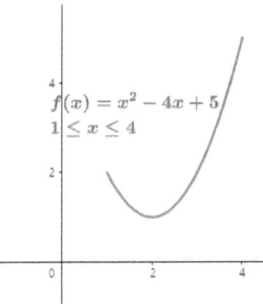

Figure 2.8a f(x).

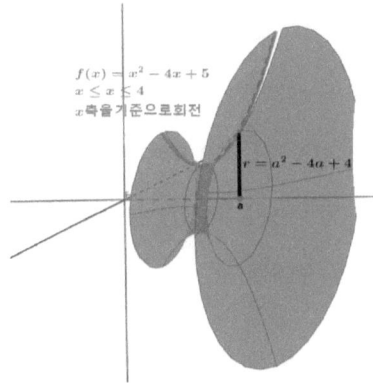

Figure 2.8b The rotation of f(x).

As shown in Figure 2.8b, the cross-sectional area of the rotating body is a circle with radius f(x). Therefore, the area of the cross-sectional area can be calculated by the function A(x). The sum of each cross-sectional area corresponding to each point of the specified section [1, 4] will be the volume of the rotating body.

That is, the volume of the rotating body is a definite integral of A(x), which is a cross-sectional function, in the interval [1, 4].

$$A(x)=\pi r^2=\pi(x^2-4x+5)^2$$
$$V=\int_1^4 A(x)\,dx=\int_1^4 (x^2-4x+5)^2\,dx$$

```
>>> x=symbols("x")
>>> f=x**2-4*x+5;f
x**2 - 4*x + 5
>>> A=pi*f**2;A
pi*(x**2 - 4*x + 5)**2
>>> v=integrate(A, (x, 1, 4));v
78*pi/5
```

Ex) Calculate the volume of the part enclosed between the two rotations when the function f(x)=x1/3 and g(x)=x/4 are rotated about the x axis, respectively.

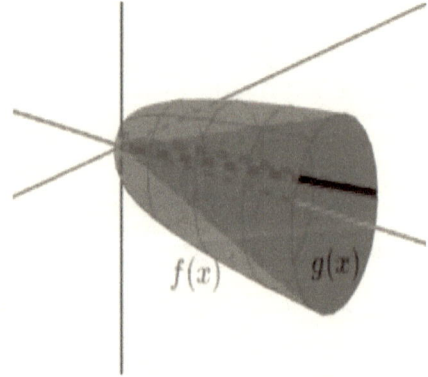

Figure 2.9 Volume contained between the two rotations.

199

From Figure 2.9,

The rotating body of f(x): green

The rotating body of g(x): red

The cross-sectional area of the portion contained between each rotation of the two functions is calculated as the function A(x). The volume of the part is the result of integrating the function A(x) in the specified section. The upper and lower bounds of the integral will be the intersection of the two functions.

$$A(x)=\pi f^2(x)-\pi g^2(x)=pi\left(x^{\frac{2}{3}}-\left(\frac{x}{4}\right)^2\right)$$

$$x^{\frac{1}{3}}-\frac{x}{4}=0 \rightarrow x=0,\ 8$$

$$\int_0^8 A(x)\,dx=\pi\int_0^8\left(x^{\frac{2}{3}}-\left(\frac{x}{4}\right)^2\right)dx=\frac{128}{5}\pi$$

```
>>> x=symbols("x")
>>> f=x**(Rational("1/3"));f
x**(1/3)
>>> g=x/4;g
x/4
>>> point=solve(f-g, x);point
[0, 8]
>>> A=pi*(f**2-g**2);A
pi*(x**(2/3) - x**2/16)
>>> v=integrate(A, (x, point[0], point[1]));v
128*pi/15
```

200

Appendix 1

for and if statements

The general characteristics of Python syntax are:

1) Python distinguishes between uppercase and lowercase letters. So 'a' and 'A' are different.

2) When using multiple statements or expressions in Python, you should add a colon(':') at the end of one expression to tell them the end of the expression.

3) If you want to add another expression in the expression, you have to insert a newline(4 spaces) after the newline.

In other words, the colon and indentation distinguish the main command expression from the subordinate command expression, and these groups are referred to as blocks.

for statement

You use a loop to simplify the use of repeated commands. The for statement has the following form:

```
for identifier in scope:
    command
```

From 'for syntax' to 'for ~ command' becomes one block. Depending on the program, you can include many blocks. In this case, the blocks represent the order in which the programs are executed. In other words, it runs from top to bottom in blocks.

In the above case, the same command is repeated until all the values of scope are passed to the identifier.

The function range() is often used to indicate a range in a 'for statement'.

```
>>> for i in range(0, 10, 2):
        print(i)
0
2
4
6
8
```

range(0, 10, 2) is an integer value in intervals of 2 from 0 to 9, so the identifier i is repeated until all the elements produced by the range() function are allocated. In order to pass values to 'i', data of all the types that can generate sequence numbers such as strings, lists, and tuple types in addition to range() is used.

The 'for loop' can be simplified by using the list comprehension syntax.

if statement

An 'if statement' is an expression for determining whether the code is executed. According to the following syntax format, 'command 1' is executed if the condition is met, otherwise 'command 2' is executed. In this syntax, the 'else ~' part can be omitted and other 'if statements' can be included in the statement.

```
If condition:
    command1
else:
    command2
```

'Command 1' shown above is included in the 'if condition' and becomes block 1. In case of 'command 2', it is included in 'else statement', so it becomes block 2. Therefore, in the above code, block 1 with 'if' is executed first, and then the execution of block 2 is decided according to the result.

Determine which values are even or not

```
>>> x=124
>>> if x % 2==0:
        print("odd")
```

'%' is an operator that returns the remainder of the division.

```
>>> 3%2
1
```

User Defined Function

To make the above code function, it starts with 'def' keyword of function generation. Create the function as follows:

```
def function name(factor(s)):
        command(s)
        return(result)
```

The 'return()' at the end of the above function is used to return the result, which means it is the end of the function. However, you can omit 'return()' if you use the 'print()' function to display the generated results in the monitor.

```
>>> def evenOddS(x):
    ....:   if x % 2==0:
    ....:       print('even')
    ....:   else:
    ....:       print('odd')
>>> evenOddS(123)
 odd
>>> evenOddS(678)
 even
```

In the above result, if the number is odd, we want to check whether it is a multiple of 3 or not. In this case, you can write the condition statement again under 'else statement'.

```
>>> def evenOddS(x):
...:     if x % 2==0:
...:         print('even')
...:     else:
...:         if x % 3==0:
...:             print('multiples of 3, odd')
...:         else:
...:             print('Not a multiple of 3, odd')
>>> evenOddS(127)
Not a multiple of 3, odd
```

Also, it is often convenient to use the keyword 'elif' statement when the 'if ~ else ~' syntax is repeated many times.

```
>>> price= 3000
>>> if price==1000:
        print("coffee")
    elif price==1500:
        print('chocolate')
    elif price==3000:
        print('juice')
    else:
        print('bonus')
juice
```

Appendix 2 Python functions

apart()

apart(x): decompose rational function x into partial fractions.

>>> f = (4*x**3+21*x**2+10*x+12) / (x**4+5*x**3+5*x**2+4*x)
>>> f
(4*x**3 + 21*x**2 + 10*x + 12)/(x**4 + 5*x**3 + 5*x**2 + 4*x)
>>> apart(f)
(2*x - 1)/(x**2 + x + 1) - 1/(x + 4) + 3/x

$$\frac{4x^3+21x^2+10x+12}{x^4+5x^3+5x^2+4x} = \frac{2x-1}{x^2+x+1} - \frac{1}{x+4} + \frac{3}{x}$$

append()

np.append (x, y): Numpy module function. create an object by adding y after the object x.

>>> a=[1,2,4,5]
>>> b=[3, 10, 4]
>>> np.append(a, b)
array([1, 2, 4, 5, 3, 10, 4])

When merging other data types, the data types are unified into one. As in the following case, the elements of the internal list are automatically converted to the character type in the list with the internal list and the character as elements. If the character and numeric types are mixed, the character type has priority.

```
>>> np.append([1,2,3],'b')
array(['1', '2', '3', 'b'], dtype='<U11')
```

arange()

np.arange(start, end, step): numpy As a module function, this function creates a sequence with the specified interval between the beginning and end.
The result produced does not include the end number.

```
>>> np.arange(10, 50, 5)
array([10, 15, 20, 25, 30, 35, 40, 45])
```

collect()

collect(eq, variable): Sorts the expression by the specified variable.

```
>>> eq=3*a*x**2 + 4*a*x - 4*a + 3*b*x**2 - 2*b*x + c*x**2 + 2*c*x;eq
3*a*x**2 + 4*a*x - 4*a + 3*b*x**2 - 2*b*x + c*x**2 + 2*c*x
```

```
>>> collect(eq,x)
-4*a + x**2*(3*a + 3*b + c) + x*(4*a - 2*b + 2*c)
```

The result sorted by collect() can be called by the object.coeff (variable, order) of each order coefficient.

```
>>> y.coeff(x, 2)
3*a + 3*b + c
>>> y.coeff(x, 1)
4*a - 2*b + 2*c
>>> y.coeff(x, 0)
-4*a
```

DataFrame()

pd.DataFrame(x, index, columns): Converts an object to a pandas object (a table type with rows and columns) as a function of the pandas module.

'index' and 'columns' are arguments for specifying row names and column names, respectively.

The type passed as an argument to this function must be a list or an array of numpy objects.

The number of lists is the number of rows, and the number of elements in the list is the number of columns.

In the following example, two lists, each consisting of three elements, are passed as arguments, resulting in a table with two rows and three columns.

```
>>> re=pd.DataFrame([["one","two","char"], [1,2,"a"]], index=['word',
'value'], columns=['first', 'second', 'third']);re
        first   second  third
key     one     two      char
value   1       2        a
```

The 'DataFrame()' function is based on the array object structure created by numpy's array() function and is a necessary data format for data analysis.

degree()

degrees(): numpy module function. convert radian to degree

```
>>> import numpy as np
>>> np.degrees(np.pi)
180.0
>>> np.degrees(np.pi/6)
29.999999999999996
>>> np.around(np.degrees(np.pi/6), 0)
30.0
```

You can use 'degrees()' function from another module 'npmath'.

```
>>> from mpmath import *
```

diff()

diff(equation, symbol, Differential count=1): Calculate the derivative of an expression or function

Derivative calculations are performed as follows.

$(x^n)'=nx^{n-1}$

The above formula is the basic algorithm of the diff() function. Therefore, the number of differentiation among the arguments of this function is to apply the above formula to the number of times. For example, the derivative for x^4 is:

$(x^4)'=4x^3$

$(4x^3)'=12x^2$

...

```
>>> diff(x**4, x)
4*x**3
>>> diff(x**4, x, 3)
24*x
>>> diff(x**4, x, x, x)
24*x
```

Instead of specifying the number of differentiation as in the above code, you can enter the variable multiple times.

doit()

object.doit(): Evaluate sympy object.

Some functions in the sympy module return only expressions as a result. This function is used to evaluate these results(objects). For example, if there is a need to preserve the integral expression itself as an object, use the integral() function instead of the integrate() function. Use doit() to evaluate the object of this expression.

```
>>> f=x**2+1;f
 x**2 + 1
>>> integrate(f, x)
 x**3/3 + x
>>> dx=Integral(f, x);dx
 Integral(x**2 + 1, x)
>>> type(dx)
 sympy.integrals.integrals.Integral
>>> dx.doit()
 x**3/3 + x
```

expand()

expand(x): Expands object (expression) x.

```
>>> expand((x+1)**2)
```

x**2 + 2*x + 1

Eq()

Eq(a, b): Returns the expression 'a = b'.

This function returns the same result as '=='. That is, the expression 'a' and the expression 'b' mean the same.

In python, '==' returns the result of 'True', 'False' as a logical expression. Also, '=' means to assign (save) the right term to the left object. Therefore, an expression such as " x-4 = 3 " should be generated as a homogeneous form like 'object = a-4-3'. A formula with one term of zero is called a homogeneous equation. In sympy, however, you can use the Eq() function to generate the following:

```
>>> solve(x-4-3, x)
[7]
>>> solve(Eq(x-4, 3), x)
[7]
```

'solve()' is a function to calculate the solution of an expression.

evalf()

object.evalf(number): Returns the effective number (n) of a 'sympy' object. Returns the same result as the function N().

>>> pi.evalf(4)
3.142.
>>> N(pi, 4)
3.142

factor()

factor(x): The object x is decomposed into elements.

>>> factor(x**2+2*x+1)
(x + 1)**2

factorial()

factorial(x): Calculates the facotial of the number x.
factorial : $n! = n \cdot (n-1) \cdot \ldots \cdot 2 \cdot 1$

>>> factorial(5) # 5·4·3·2·1
120

fraction()

fraction(x): Returns the numerator and denominator of the fractional x, separately.

```
>>> f=fraction((x-1)/(x**2-2*x+1));f
 (x - 1, x**2 - 2*x + 1)
>>> f[0]
 x - 1
>>> f[1]
 x**2 - 2*x + 1
```

Function()

Function(name of function)(symbol): Creates a function for the specified variable. If you do not specify a variable, simply create a function with the specified function name.

```
>>> x, y=symbols('x, y')
>>> f=Function('f')(x);f
 f(x)
>>> diff(f, x)
 Derivative(f(x), x)
>>> f.diff(x)
 Derivative(f(x), x)
>>> g=Function('g');g
 g
```

```
>>> g.diff(y)
 1
>>> g.diff(x)
 1
>>> g=Function('g')(x);g
 g(x)
>>> g.diff(x)
 Derivative(g(x), x)
>>> g.diff(y)
 0
```

Hyperbolic Function

The hyperbolic functions are represented as sinh(). Sympy modules can also be calculated using functions with the same name.

sinh(x), cosh(x), tanh(x), csch(x), sechx(x), coth(x)

The inverse of the hyperbolic function is used with the prefix 'a-' for each function. For example, sinh() is as follows.

```
>>> x=symbols('x')
>>> sinh(x)
 sinh(x)
>>> N(sinh(1), 3)
 1.18
>>> f=(exp(x)-exp(-x))/2;f
```

```
exp(x)/2 - exp(-x)/2
>>> N(f.subs(x, 1), 3)
1.18
>>> N(asinh(sinh(1)), 3)
1.00
```

integrate()

A sympy module function that performs integral calculations of expressions.

indefinite integral : integrate(equation, symbol)

definite integral : integrate(equation, (symbol, low, upper))

Note that in sympy the calculation of the indefinite integral is omitted from the constant term.

```
>>> f=3*x**2+2*x+1;f
3*x**2 + 2*x + 1
>>> integrate(f, x)
x**3 + x**2 + x
>>> integrate(f, (x, -1, 3))
40
```

limit()

limit(f, var, val, dir): Returns the limit result for the object (expression, f). A mathematical representation of this function is as follows:

$$\lim_{x \to a^d} f$$

In order for a function to have a limit value in 'a', the value should be the same when approaching left and right around 'a'. 'd' specifies the direction of the limit. Access from the left is '-' (left limit) and access from the right is '+' (right limit).

The following is the calculation of the right and left limits of the function. This function is discontinuous at x = 0 because the right and left limits are different. That is, there is no limit at x = 0.

```
>>> x=symbols("x", real=True)
>>> f=1/x
>>> limit(f, x, 0,"+")
  oo
>>> limit(f, x, 0, '-')
  -oo
>>> limit(f, x, 0)
  oo
```

linspace()

np.linspace(start, end, size): Function of the numpy module. Creates a sequence of the specified size at regular intervals from start to end.

```
>>> import numpy as np
 >>> np.linspace(1, 10, 7)
array([ 1. , 2.5, 4. , 5.5, 7. , 8.5, 10. ])
```

list.append()

object.append(): List method that adds a new element to the end of the list object

```
>>> a=[1,2,3]
>>> a.append(4)
>>> a
 [1, 2, 3, 4]
```

list comprehension

Loops can be created simply by using the list comprehension syntax, and the result is returned as a list.

object=[command for loop [if condition]]

In this syntax, you can add conditional statements after the loop statement.

```
>>> x=[2, 4, 7]
>>> y=[3, 5, 6]
>>> x+y
>>> [2, 4, 7, 3, 5, 6]
>>> z=[x[i]+y[i] for i in range(len(x))]
>>> z
>>> [5, 9, 13]
```

log()

log(x, b=None): A sympy module function that computes the log function. If the base b is not specified, it means natural log (base = e)

If a base number is specified, it can be transformed into a natural log using the properties of the log and calculated as follows:

$\log_b x = \log(x)/\log(b)$

$y=\log_2 16 \rightarrow 2y=16 \rightarrow y=4$

```
>>> x=symbols("x")
>>> log(16, 2)
4
>>> N((log(16))/(log(2)), 1)
4.
```

$y = \log 3/227/8 \rightarrow (3/2)y = 27/8 \rightarrow y = 3$

```
>>> N(log(27/8, 3/2), 2)
3.0
```

min(), max()

min(x): The python intrinsic function returns the minimum value in a continuous-number object (x), such as a list or tuple.

max(x): The python intrinsic function returns the maximum value within a continuous-number object (x), such as a list or tuple.

```
>>> min(1,4,9)
1
>>> max(1,4,9)
9
```

N()

N(x, n): The function indicates any number(x) to the specified number(n) of significant digits

This function returns the same result as the 'object.evalf(n)', but the N() function also works for a common numeric type, not a sympy object.

```
>>> N(1.34256, 4)
```

nsimplify()

nsimplify(expression, rational=True):   This function makes the coefficient(s) in an expression irreducible fraction(s)

"rational == True" is the default. The coefficients of the equation are returned in fraction form.

  "tolerance" is an estimate of floating number. The smaller the value, the greater the accuracy. The default is None.

np.pi: The Attribute of the numpy module that generates π

```
>>> np.pi
3.141592653589793
>>> nsimplify(np.pi)
314159265358979/100000000000000
>>> nsimplify(np.pi, rational=True)
314159265358979/100000000000000
>>> nsimplify(np.pi, rational=True, tolerance=0.1)
22/7
>>> nsimplify(np.pi, rational=True, tolerance=0.001)
355/113
```

Piecewise()

Piecewise((expr1, cond1),(expr2, cond2),...): Calculate the piecewise function. That is, it performs the appropriate calculation according to the conditions of the expression.

In the following example, the function is defined under the condition of variable x>0. Therefore, the function specifies a value of 0 in the range of x≤0.

$$f(x)=\begin{cases} \log(x) & \text{if } x>0 \\ 0 & \text{if } x\leq 0 \end{cases}$$

```
>>> f = log(x)
>>> p = Piecewise( (0, x<0), (f, x> 0))
>>> p.subs(x,100)
log(100)
>>> p.subs(x,-100)
0
```

radians()

radians(x): numpy module function. degree (angle) converted to radian.

```
>>> np.radians(60)
1.0471975511965976 #=π/3
>>> N(pi/3)
```

1.04719755119660

You can use 'radians ()' in another module 'npmath'.

```
>>> from mpmath import *
>>> x1=radians(270);N(x1,4)
4.712
```

range()

range(start, stop, step): Generates sequences at regular intervals up to the specified start and end.

1) This function works only on integers.

2) The beginning of arguments can be omitted. The default is 0.

3) The generated sequence does not include the end number.

4) To specify a step, you must enter the argument 'start'.

5) You can create a sequence that is decremented by negative 'step'.

```
>>> list(range(1, 10, 2))
 [1, 3, 5, 7, 9]
>>> list(range(10))
 [0, 1, 2, 3, 4, 5, 6, 7, 8, 9]
>>> list(range(20, 10, -2))
 [20, 18, 16, 14, 12]
```

list(x): a function to convert an object x to a list type.

Rational()

Rational(x) or Rational(nominator, denominator): Keeps the form of fraction x. Generally, the fractional form is calculated and represented as a decimal number, but if you specify a fraction using this function, the form is retained during the calculation, so the result is also returned in the form of a rational expression.

```
>>> Rational('5/3') #=Rational(5, 3)
5/3
>>> 5/3
1.6666666666666667
>>> 2+Rational('5/3')
11/3
>>> 2+5/3
3.66666666666667
>>> 2+Rational(5, 3)
11/3
```

sign()

sign(x): Returns the sign of the expression x. For example, if f(x) = |x| is differentiated, f '(x) = 1 or -1.

$$f(x)=\begin{cases} x \text{ if } x>0 \\ 0 \text{ if } x=0 \\ -x \text{ if } x<0 \end{cases} \rightarrow f'(x)=\begin{cases} 1 \text{ if } x>0 \\ 0 \text{ if } x=0 \\ -1 \text{ if } x<0 \end{cases}$$

However, the results returned by diff() are:

```
>>> x=symbols('x', real=True)
>>> f=abs(x);f
 Abs(x)
>>> dx=diff(f, x);dx
 sign(x)
```

The meaning of sign(x) in the above result is as follows.

$x>0 : |x|=x \rightarrow sign =1$

$x<0 : |x|=-x \rightarrow sign=-1$

$x=0 : |x|=0 \rightarrow sign=0$

Thus, sing (x) means that the result depends on x.

The results of f '(2) and f (- 2) in the above example f(x) = |x| are

as follows.

```
>>> dx.subs(x, 2) #+1×1
 1
>>> dx.subs(x, 0)#0×1=0
 0
>>> dx.subs(x, -2)#-1×-1=1
 -1
In case of f(x)=x³,
>>> x=symbols('x', real=True)
>>> f=abs(x**3)
x**2*Abs(x)
>>> df=diff(f, x);df
x**2*sign(x) + 2*x*Abs(x)
```

```
>>> df.subs(x, -3)
-27
>>> df.subs(x,3)
27
```

simplify()

simplify(x): Converts the expression x(object) to a simple form.

```
>>> a = (x + x**2)/(x*sin(t)**2 + x*cos(t)**2);a
(x**2 + x)/(x*sin(t)**2 + x*cos(t)**2)
>>> simplify(a)
x + 1
```

The above equation is as follows.

$$a = \frac{x+x^2}{x\sin^2(t)+x\cos^2(t)} = \frac{x(x+1)}{x(\sin^2(t)+\cos^2(t))} = x+1$$

solve()

solve(function or expression, symbol): Calculates the solution of the equation. The result is returned in list form.

For example, the solution of 3x-6 = 0 is 2. This is coded as follows:

```
>>> x=symbols("x")
>>> solve(3*x-6, x)
```

This function computes the solution to the multivariate of several expressions, such as the simultaneous equations. In this case, the result returns the solution corresponding to the variable in dictionary form.

 solve([equation1, equation2,...], (symbol1, symbol2...))

The expressions in the function's arguments are passed in list form, and the variables in tuple form. You do not have to specify a variable.

For example,

$2x+3y=6$

$4x+y=10$

```
>>> x, y=symbols("x, y")
>>> eq1=2*x+3*y-6;eq1
2*x + 3*y - 6
>>> eq2=4*x+y-10;eq2
4*x + y - 10
>>> solve([eq1, eq2], (x, y))
{x: 12/5, y: 2/5}
```

Two methods can be applied to generate the appropriate expressions to pass to solve(). Create an object called eq1, eq2 (stored in the repository) as above.

The above expression is a form of 2x + 3y-6 = 0, and one of the terms of the expression is 0. This form is called a homogeneous equation.

In addition to the form of the homogeneous equation, we use the 'Eq()' function to generate and apply each expression.

```
>>> eq3=Eq(2*x+3*y, 6);eq1
Eq(2*x + 3*y, 6)
>>> eq4=Eq(4*x+y, 10);eq2
Eq(4*x + y, 10)
>>> solve([eq3, eq4], (x, y))
{x: 12/5, y: 2/5}
```

You can also use booleans, such as inequalities.

```
>>> solve(x**2-1, x)
[-1, 1]
```

The above code specifies a variable for the solution of the expression, but the specification of the variable is not required. You can execute it without assigning variables as shown in the following example.

```
>>> solve(x<3)
(-oo < x) & (x < 3)
```

In the following expression, sin(0) is 0 at x=0, but this expression is undefined because the denominator is zero. Therefore, solutions must not contain zeros.

sin(x)/x=0

```
>>> solve(sin(x)/x)
[pi]
```

However, to ignore these conditions and return 0, specify 'check' in the function's arguments as False.

```
>>> solve(sin(x)/x, check=False)
[0, pi]
```

Thus, the selectivity of the solve() function is very variable. Typically, 'dict=True' represents the solution in dictionary form by mapping each variable and value, and 'set=True' represents the variable in list format and solution in the form of set of tuples.

```
>>> solve(x**2-1, x, dict=True)
[{x: -1}, {x: 1}]
>>> solve(x**2-1, x, set=True)
([x], {(-1,), (1,)})
```

solve_undermined_coeffs()

solve_undermined_coeffs(expression created by Eq(), [Coefficients], variable): Calculates each coefficient contained in an expression. The result is returned in dictionary format.
For example, the coefficients(a, b) of 2ax+a+b=x,
2a=1 →a=1/2, b=-1/2

```
>>> x, a, b=symbols("x, a, b")
>>> solve_undetermined_coeffs(Eq(2*a*x+a+b, x), [a, b], x)
{a: 1/2, b: -1/2}
```

sort()

sort(x): numpy Modular function that sorts x in ascending order.

```
>>> a=array([55, 75, 55, 73, 53, 42, 2, 28, 37, 42])
>>> np.sort(a)
array([ 2, 28, 37, 42, 42, 53, 55, 55, 73, 75])
```

subs()

object.subs(symbol, value or other symbol): Computes by substituting the value or the new variable(s) for the specified variable(symbol) in the object. Of course, this is an object created using the sympy module. Also, if there are several variables to be replaced, the argument is passed in dictionary format.

```
>>> x, y, x1, y1=symbols("x, y, x1, y1")
>>> f=x+2*x**2+y;f
2*x**2 + x + y
>>> f.subs(x, 3)
y + 21
>>> f.subs({x:3, y:1})
22
>>> f.subs({x:x1, y:y1})
2*x1**2 + x1 + y1
```

sum()

A built-in function in python that computes the sum of objects x constructed from numbers.

Built-in functions are core functions that do not need to be called from outside.

```
>>> a=[3,2,1,4]
>>> sum(a)
10
```

Symbols specified in the sympy object are calculated by the sum() function, because they can be replaced by numbers.

```
>>> A,B,C,D,E=symbols('A,B,C,D,E')
```

```
>>> sum((A, B, C, D, E))
 A + B + C + D + E
>>> a.subs(A, 3)
 B + C + D + E + 3
>>> a.subs({A:3, B:7})
 C + D + E + 10
```

Sum()

Sum(f, (var, dw, up)): This function performs the sum from the lower limit (dw) to the upper limit (up) for the expression (f).

$$\sum_{var=dw}^{up} f$$

var: variable

This function returns a series expression (sympy object) instead of returning the result of the direct computation. Therefore, it is used with the doit () function to calculate this result object.

```
>>> total=Sum(x**2, (x,0, 3));total
 Sum(x**2, (x, 0, 3))
>>> total.doit()
 14
```

symbols()

symbols('What to use as a symbol', ...): Specify 'what' as a symbol(variable). The 'var()' function is equivalent.

A function can specify multiple symbols at the same time.

In general, when you specify a symbol, the range of the symbol(variable) includes all areas such as real numbers, complex numbers, and so on. However, there are cases where you have to limit the area to 'real' in your calculations. In this case, specify the area as follows.

```
>>> x, u=symbols("x, u")
```

The character or string specified by the symbol can be operated.

```
>>> f=x**2+2*x+1; f
 x**2 + 2*x + 1
>>> f1=3*x+3;f1
 3*x + 3
>>> f+f1
 x**2 + 5*x + 4
```

In the above case, the range of symbols x and u includes all real and complex numbers. If you want to define a positive variable x, pass 'positive = True' as an argument as follows:

```
>>> x=symbols("x", positive=True)
>>> x>0
```

True

>>> x<0

False

sympify()

sympify(x): Convert object x to a sympy object

```
>>> a=3.14
>>> type(a)
 float
>>> a1=sympify(a);a1
 3.14000000000000
>>> type(a1)
 sympy.core.numbers.Float
```

together()

together(x): The function makes a fraction reduction to common denominator. However, the results of this function are not simplified. Therefore, this result can be simplified by the simplify() function.

```
>>> f1=apart(f);f1
 (2*x - 1)/(x**2 + x + 1) - 1/(x + 4) + 3/x
>>> f2=together(f1);f2
```

$$(x*(x + 4)*(2*x - 1) - x*(x**2 + x + 1) + 3*(x + 4)*(x**2 + x + 1))/(x*(x + 4)*(x**2 + x + 1))$$

The result of the together() function can represent the numerator and denominator of a fraction using the numer() and denor() functions, respectively.

numer(fraction): Represents a numerator.

denom(fraction): Denominator.

```
>>> numer(f2)
x*(x + 4)*(2*x - 1) - x*(x**2 + x + 1) + 3*(x + 4)*(x**2 + x + 1)
>>> denom(f2)
x*(x + 4)*(x**2 + x + 1)
```

trigonometric functions

The sympy functions for the trigonometric functions are:

sin(), cos(), tan(), csc(), sec(), cot()

Arguments for these functions must be passed in radians.

```
>>> x=pi/3
>>> sin(x)
 sqrt(3)/2
>>> cos(x)
 1/2
>>> tan(x)
 sqrt(3)
```

```
>>> csc(x)
 2*sqrt(3)/3
>>> sec(x)
 2
>>> cot(x)
 sqrt(3)/3
```

For the inverse trigonometric function, use the prefix 'a' for the above trigonometric function.

$asin(x) = sin^{-1}(x)$ $acsc(x) = csc^{-1}(x)$

$acos(x) = cos^{-1}(x)$ $asec(x) = sec^{-1}(x)$

$atan(x) = tan^{-1}(x)$ $acot(x) = cot^{-1}(x)$

```
>>> x=1
>>> asin(x)
 pi/2
>>> acos(x)
 0
>>> atan(x)
 pi/4
>>> acsc(x)
 pi/2
>>> asec(x)
 0
>>> acot(x)
 pi/4
```

INDEX

www.ingramcontent.com/pod-product-compliance
Lightning Source LLC
Chambersburg PA
CBHW021812170526
45157CB00007B/2559

9781097682799